Information Circular 9468

Developing Random Virtual Human Motions and Risky Work Behaviors for Studying Anthropotechnical Systems

By Dean H. Ambrose

U.S. DEPARTMENT OF HEALTH AND HUMAN SERVICES
Centers for Disease Control and Prevention
National Institute for Occupational Safety and Health
Pittsburgh Research Laboratory
Pittsburgh, PA

March 2004

ORDERING INFORMATION

Copies of National Institute for Occupational Safety and Health (NIOSH)
documents and information
about occupational safety and health are available from

NIOSH–Publications Dissemination
4676 Columbia Parkway
Cincinnati, OH 45226-1998

FAX: 513-533-8573
Telephone: 1-800-35-NIOSH
(1-800-356-4674)
E-mail: pubstaff@cdc.gov
Web site: www.cdc.gov/niosh

This document is the public domain and may be freely copied or reprinted.

Disclaimer: Mention of any company or product does not constitute endorsement by NIOSH.

DHHS (NIOSH) Publication No. 2004-130

CONTENTS

	Page
Abstract	1
Introduction	2
Background	2
Random motion application	3
Developing random motion code	3
Random motion evaluation	4
Random virtual human motion	6
Random motion rules	8
Random behavior motion	8
Virtual human attributes: viewing area	11
Calculations and output	12
Conclusions	13
Acknowledgments	14
References	14
Appendix A.—Code samples of random virtual human motion scenarios of behavior	15
Appendix B.—Code samples of calculations and output for the random virtual human motion code	24

ILLUSTRATIONS

1.	Operator contacted in the hand (or fingers) and leg	2
2.	Motion path of cup while drinking from it	3
3.	Virtual human figure in Jack	4
4.	Distance traveled between the cube and a reference point	5
5.	Start-to-finish hand motions	6
6.	Full-scale roof bolter assembly setup for data collection	7
7.	Virtual operator's behavior hand on drill bit	9
8.	Virtual operator's behavior hand on boom arm	9
9.	Virtual operator's behavior hand off drill and boom arm	9
10.	Angular data of the original and modified viewing areas for the virtual operator	11
11.	Reference points on the viewing cone and boom arm	12

TABLES

1.	Random motion rules for a roof bolter machine	8
2.	Work behavior list for drilling the hole and installing a bolt	9
3.	Decision logic for the algorithm to determine behavior selection	10
4.	Sample data output file	13

UNIT OF MEASURE ABBREVIATIONS USED IN THIS REPORT

cm	centimeter	in	inch
cm/sec	centimeter per second	in/sec	inch per second
fL	footlambert	sec	second

DEVELOPING RANDOM VIRTUAL HUMAN MOTIONS AND RISKY WORK BEHAVIORS FOR STUDYING ANTHROPOTECHNICAL SYSTEMS

By Dean H. Ambrose[1]

ABSTRACT

A computer model was created that generates contact data by means of simulation while altering several variables associated with the machine and its operator. These variables include work environment, the operator's anthropometry, work posture, choice of risky work behavior, and the machine's appendage velocity. In the model, a contact means two or more objects intersecting or touching each other, e.g., the appendage colliding with the operator's hand, arm, head, or leg. This report documents the code development of special features of the computer model, random virtual human motions and behaviors, which made it possible for researchers to study hazardous interactions, such as contacts between the operator and machine. The original idea for random virtual human motions began with the need for a model that, when simulated, would experimentally mimic machine and human actions that could cause actual injuries or fatalities in the workplace. Injury incident investigation reports do not usually contain enough information to aid in studying this problem, and lab experiments with human subjects are also not feasible because of safety issues and ethical issues.

Early investigation that evaluated the feasibility of random motions resulted in a primary programming structure and building blocks for code expansion. Basic motions and standard derivation data provided random parameters for the operator's movement in the random motion code. Experimenting with basic motions in preliminary models helped researchers devise a random motion technique that worked with xyz motion parameter values from basic motion parameters. Both values were the building blocks used in motion commands that comprised the core of the random motion code. When the code is executed in a three-dimensional computer model, it generates realistic starting positions for the operator's (virtual human) body and body parts and realistic motions mimicking a work sequence with a valid range of variation in movements. Researchers developed several rules for the random motion technique that, when applied, will maintain directional integrity of a virtual human's basic motions. From previous research, researchers formulated the algorithm that determines the behavior during machine operation. The decision algorithm was needed by the random motion code to randomly select what behavior to use for a simulation execution for a realistic representation of the operator's motions and risky behaviors during work tasks. The operator's chance of avoiding the moving appendage was also a concern to ensure that data reflecting a near-miss would not be considered a contact. Therefore, to filter a database obtained from simulations for actual contacts, it was necessary to track the operator's field of view and determine from this information when the appendage is in and out of the operator's view.

NIOSH researchers successfully applied random motions and risky behaviors to examine the speed range of a roof bolter machine appendage for different workplace scenarios and compare statistically which is most likely to cause contacts (and possible injuries) to miners. Researchers in this study believe that the use of such simulations, treated with advanced statistical procedures, are extremely useful tools to evaluate the hazards of tasks where it is not possible to perform experiments with human subjects.

---

[1] Safety engineer, Pittsburgh Research Laboratory, National Institute for Occupational Safety and Health, Pittsburgh, PA.

INTRODUCTION

Several injuries to operators of underground coal mining equipment have led to an investigation of safe vertical velocities of a roof bolter boom arm at the National Institute for Occupational Safety and Health (NIOSH)'s Pittsburgh Research Laboratory. There are currently no regulations or method of determining the safe speed of roof bolter boom arms. Accident investigation reports from the Mine Safety and Health Administration (MSHA) do not usually contain enough information to aid in studying interactions between a machine and its operator. Lab experiments with human subjects also are not feasible because of safety and ethical issues. With this in mind, NIOSH researchers successfully developed a computer model that uses simulation software. The model generates contact data by means of simulation while exercising the model with several variables associated with the machine and its operator. These variables include coal seam height, the operator's anthropometry, work posture, choice of risky work behavior, and the machine's appendage velocity. In the model, a contact means two or more objects intersecting or touching each other, e.g., the boom arm colliding with the operator's hand, arm, head, or leg (figure 1). This report documents the development of special features of the computer model, random virtual human motions and behaviors, which made it possible for researchers to study hazardous interactions between the operator and machine.

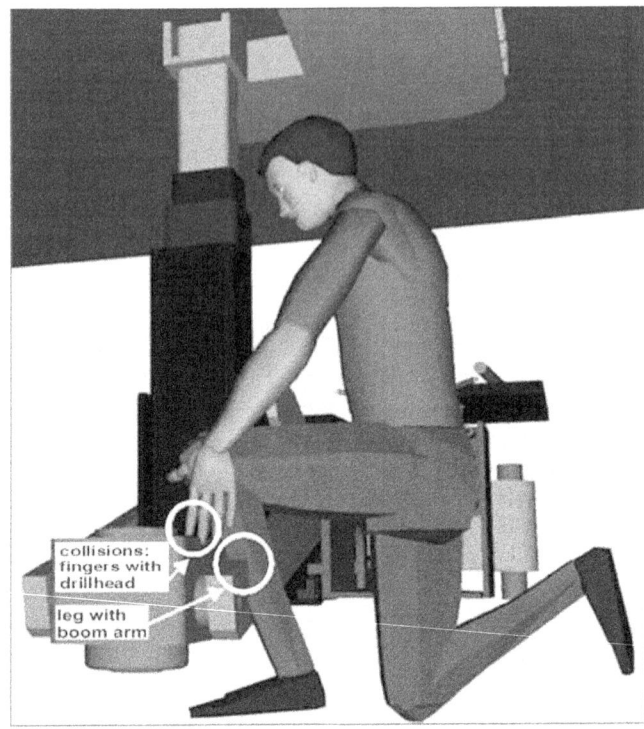

Figure 1.–Operator contacted in the hand (or fingers) and leg.

BACKGROUND

To most people, the word "model" evokes images of miniature aircraft in wind tunnels or cockpits disconnected from their airplanes to be used as pilot training. It has been found useful to build physical models to study anthropotechnical systems. Anthropometry is the science of human body measurement. An anthropotechnical system is defined as an operator, together with the machine he or she operates, and, in the context of this report, a virtual system. This report addresses a mining machine known as a roof bolter and its operator.

The original idea for random virtual human motions began with the need for a model that, when simulated, would experimentally mimic machine and human actions that could cause actual injuries or fatalities in the workplace. Ambrose [2000] reported that simulation executions using the model would allow researchers to generate, collect, and analyze contact data within the motions and events from the virtual system. The database of contacts could then be used to study motions and events from the operator and machine movements, for example, one could examine what range of appendage speeds minimizes the virtual operator's chances of making contact with the moving appendage.

One of the most difficult problems faced by a model that generates human motions is accurately representing the actual virtual system being studied. The uncertainty and randomness inherent in an operator's tasks can be compared to someone drinking a beverage from a cup. The occurrence of lifting the cup to one's mouth and placing it back onto the table is considered a random motion, and one could easily visualize the path of that motion (figure 2). To model this random motion, the sequence of someone drinking a beverage from a cup would recur until the cup is empty. Each motion path would differ slightly even though the motions basically look alike. A model incorporates the randomness of the motion and path variance within that motion. Likewise, in the case of a machine operator, the operator's behaviors, motions of each behavior, and motion paths associated with each motion behavior will have some degree of randomness. This factor of randomness gives the model a realistic representation of the operator's motions and behaviors during any task.

Finding the right software tool with all the features and capabilities to develop and execute computer code for virtual human motions and behaviors was critical. Software tools

Figure 2.—Motion path of cup while drinking from it.

such as Anthropos, Jack, Ramsis, and Safework all provided a commercial product with a virtual human modeling system for ergonomic analyses and work performance evaluations to help design and study anthropotechnical systems. UGS PLM Solutions' Jack version 1.2 was chosen as the software tool to develop the random virtual human motions and behaviors for use in roof bolter models because it allows users without source code to extend or modify Jack.

Jack is an ergonomics modeling and analysis software application that tests human integration of machine designs and evaluates the safety of workplace design through special toolkits for any work environment and job postures. In addition, the software includes a proven anthropometry scaling system [Grosso et al. 1989; U.S. Army 1988; NASA 1987; Pandya 1992a,b]. Jack easily imports CAD models, such as a roof bolter accurately modeled in Pro/Engineering software, into a simulation environment. An evaluation of Jack's human-centric visual simulation software package revealed its potential as a core software platform to support human factors research. Furthermore, the software architecture allows users without source code to extend or modify Jack's simulation functionality by writing code with the LISP programming interface and Jack command language (JCL). Jack's software architecture gave researchers the means to develop and execute computer code that generates random virtual human motions and risky behaviors associated with machine tasks.

RANDOM MOTION APPLICATION

NIOSH successfully applied random motions and risky behaviors to examine the speed range of a machine appendage for different workplace scenarios and to compare statistically which is most likely to cause contacts (and possible injuries) to miners. Ambrose et al. [forthcoming] report in detail the investigation of safe vertical velocities of a roof bolter boom arm. Results of the data analysis of roof bolter simulations show that the virtual operator's response time has little effect on the number of contacts experienced. Based on frequency and cross-tabulation analyses, regardless of other variables, contact incidents were always greater when the bolter boom arm was moving up, were always greater on the hand, and were always greater for the boom arm part of the machine. Also, the 25th-percentile-sized operators experienced more contacts than other operator sizes and had most of their contacts during speed 33 cm/sec (13 in/sec). The 152-cm (60-in) mine seam experienced more contacts than other seam heights tested and had most of the contacts during speed 41 cm/sec (16 in/sec). Results of a survival analytic approach suggest that the speed of the boom arm is the most important factor in determining the risk of an operator making contact. Based on the data collected, boom arm speed greater than 41 cm/sec (16 in/sec) results in a substantial increase in risk of contact to the roof bolter operator. Speeds less than or equal to 33 cm/sec (13 in/sec) are associated with a more modest relative risk, which represents an acceptable level of risk. Analysis information is helpful to the mining industry in making recommendations that reduce the likelihood that roof bolter operators will experience injury due to contact with a moving boom arm.

DEVELOPING RANDOM MOTION CODE

Understanding Jack's human motion kinematics and motion system was essential to random motion and behavior development. The virtual human figure in Jack (figure 3) is represented as a Peabody figure, which is a collection of segments connected by joints. In addition, Jack human figures are controlled through a set of kinematics constraints that make the figure behave in a certain way. These constraints are maintained automatically by Jack software so they are transparent to the user. Jack's human motion kinematics are well defined and validated [Azuola et al. 1996].

Jack's manipulation process defines how the virtual human is to achieve the final posture for the whole body, head, back, hand, arm, or leg. The motion that the virtual human goes through to achieve a final posture is described only through Jack's motion system by interpreting motion commands created for the figure by the end-user (researcher). In this motion system, each motion is active over a specific interval of time, delimited by a start time and an ending time. Thus, the motion system simulates a model having elements of continuous motion. However, each motion creation

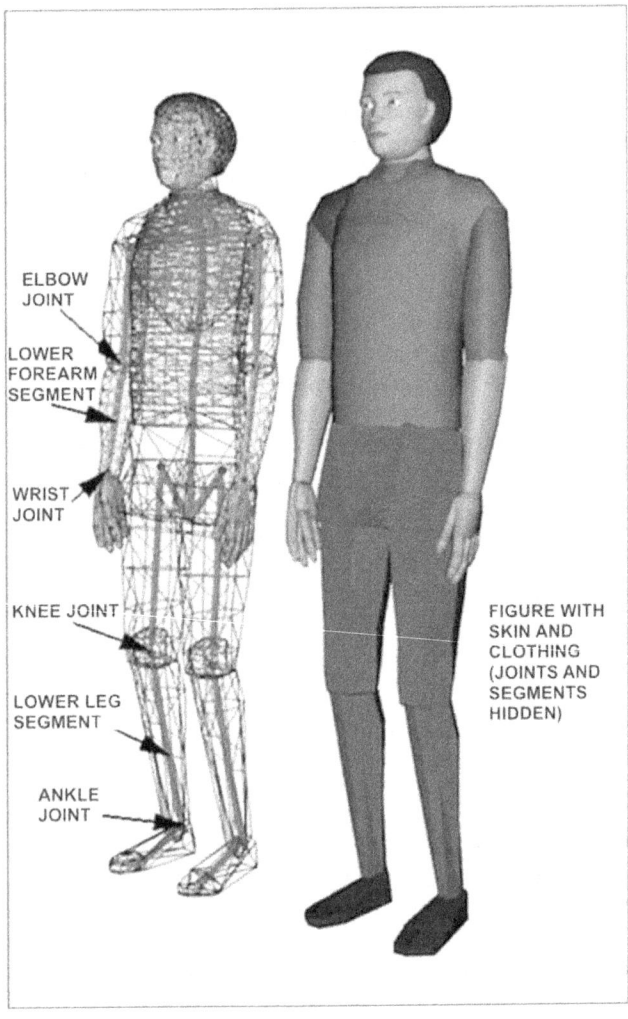

Figure 3.—Virtual human figure in Jack.

posture within a specific interval of time. Therefore, the system simulates a model having elements of discrete-event motion. Because Jack's motion system does not simulate simultaneously completely discrete-event or completely continuous motions, researchers needed a model with aspects of both motions. NIOSH researchers developed a unique code that combines discrete-event and continuous simulation with built-in capability of producing random manipulation values before transforming them into a motion path by the motion system. Subsequently, Jack's motion system reflects the variance in that motion path of the virtual human as defined by the motion parameters.

The following sections of this report discuss the various development activities regarding random virtual human motion and risky work behavior. The discussions include (1) the early investigation that evaluated the feasibility of random motions, (2) stages of developing and validating random virtual human motion models, (3) rules to help apply random virtual human motions to specific machine models, (4) random behavior that defines the virtual human motions when operating a specific machine, e.g., roof bolter, (5) the optimal viewing area of the virtual machine operator and how it contributes to when and if a contact occurs, and (6) the information from the model that is calculated and printed to a file.

RANDOM MOTION EVALUATION

Initial test trials on random motion were made easy for researchers because the Common LISP language interface to Jack allows general programming of Jack internals, which simplifies code development. Early in program code development, researchers successfully tested code from LISP and JCL that generated random movements of a cube object within the motion system. Only manipulation values for xyz-positional coordinates of the cube were tested. Researchers accepted the premise that if one set of manipulation values works, so will the other. In addition, the Common LISP facility for generating pseudorandom numbers has been carefully defined by LISP developers to make its use portable. The function **random** accepts a positive number (N) and returns a number of the same kinds between zero (inclusive) and N (exclusive). An approximately uniform choice distribution is used [Steele 1990]. When executed, the following portion of the JCL and LISP programming code generates random movement of a cube object:

command has values (in the context of this report, "manipulation values") for each of the motion parameters. The software automatically generates manipulation values when the user interacts with the software to manipulate a desired posture or position of the virtual human. For example, generated manipulation values for xyz-orientation angles and xyz-positional coordinates define the final posture position of the virtual human.

The human motion's algorithm generates and animates in Jack's motion system the motion path to achieve this final

```
.....setq nn 0                                          ;; sets nn to 0
(dotimes (k 50 t)                                       ;; counter
(jcl 'move_figure (js "cube")                           ;; jcl command moves cube
(jtrans (-100 (random 200)) (+ 100 (random 100)) (- 100 (random 200)))) ;; random motion values
(redraw)                                                ;; readies for the next motion
(setq r (global_position "refpt" "refpt.base")          ;; defines reference point
(setq c (global_position "cube" "cube.base")            ;; defines the point on a cube
(incf nn) .....                                         ;; increases counter nn by 1
```

The xyz-positional coordinates of the cube object are represented by (jtrans (- x (random 200))(+ y (random 100))(- z (random 200)). The x, y, and z values are each 100. xyz's positive numbers (N) for the random function are 200, 100, and 200, respectively. All values were chosen to cause a large displacement in the cube's movements. The following equation represents the manipulation values of the cube produced by the random motion technique using just positional "x" value:

$$\{positional \ldots X\} \rightarrow (N_0) - [a \text{ random number from zero to } N], \tag{1}$$

where N_0 is the original X value from a motion scenario, and

N is a variable that sets the upper limits of the random number range.

Calculations and output, other important sections of the random motion code, include random distance between a point on the cube (c) and a reference point (r) during 50 iterations (see figure 4). The calculated distance, D, between these points is given by the formula:

$$D = \sqrt{(X_c - X_r)^2 + (Y_c - Y_r)^2 + (Z_c - Z_r)^2} \tag{2}$$

The final version of the random motion code had to have the capability to perform calculations and record the data to a computer file, because the primary feature for the roof bolter model was to generate and collect data. The following portion of the JCL and LISP code shows this capability by calculating changes in distance of the cube and output the results to file:

```
....(format outfile "~A" nn)                    ;; outputs iteration number to a file
(format outfile " ~A\n"                         ;; calculates distance and outputs to a file
(sqrt (+ (* (- (aref c 0) (aref r 0))(- (aref c 0) (aref r 0)))
(* (- (aref c 1) (aref r 1))(- (aref c 1) (aref r 1)))
(* (- (aref c 2) (aref r 2))(- (aref c 2) (aref r 2))))))  ....
```

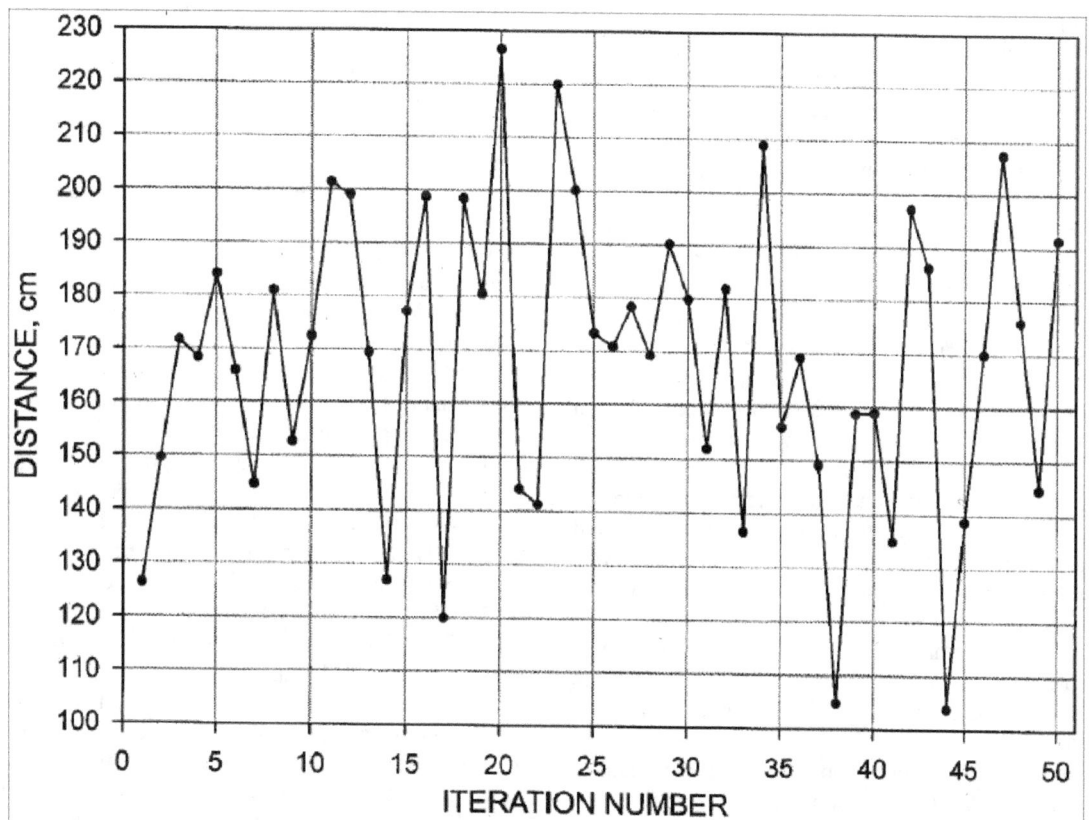

Figure 4.—Distance traveled between the cube and a reference point.

Successfully tested random motion within Jack's motion system was an important step toward its use on virtual human motions. Testing resulted in a primary programming structure and building blocks for code expansion that would help researchers develop random movement in objects and calculate and store desired information on the action.

RANDOM VIRTUAL HUMAN MOTION

A virtual human motion is a movement of a part of the body from one position to another. The body movement of how to get there is specified in terms of the final positions and parameters such as manipulation values. When motion starts in any part of the body, it implicitly defines the initial position of the motion. For example, the sequences of movements for a hand are defined with one motion for each hand movement having one set of manipulation values. Each motion serves to move the hand from its current position when the motion starts, wherever that may be, to the final position for that motion (figure 5). Therefore, it is important that researchers have correct human motion data to study. For example, in the case of incorporating random motion into virtual human motion of the roof bolter model, movements of the operator working the machine were needed. Video footage from roof bolter operations in an actual underground coal mine and manufacturer's training videos were used to help develop basic motions for preliminary models. Regarding roof bolting operations, a preliminary model has basic motions of human (operator) movements that do not touch the machine's moving appendage. Thus, developers took extra care to ensure that basic motions from the virtual operator and machine did not induce contacts. These motions from the preliminary models provided the original manipulation values (xyz values) for each of the motion parameters. These critical values, manipulation values for xyz-orientation angles and xyz-positional coordinates, formed random motions to help build motion commands for the code.

The study by Bartels et al. [2001] of human motion envelops ensured that parameter assumptions made for the basic motions conform to actual field practice. Human subject tests with a full-scale wooden working mockup of a roof bolter boom arm and a motion tracking system were used to collect human motion data and to check input motion parameters (figure 6). The results were compared to lab and computer simulation results. Motion variance analysis showed that the range of variation (2 to 30 cm) is within normal parameters of human motion. When reviewing the data as a function of anthropometrical scale, the variation in movement produces a consistent pattern, and researchers calculated a standard deviation for these points for each anthropometry.

Researchers also used the human motion envelop data to calculate an average starting position measured from the back on a subject. All subjects were measured as a function of different mine environment scenarios, and standard deviations for these points were determined. Researchers

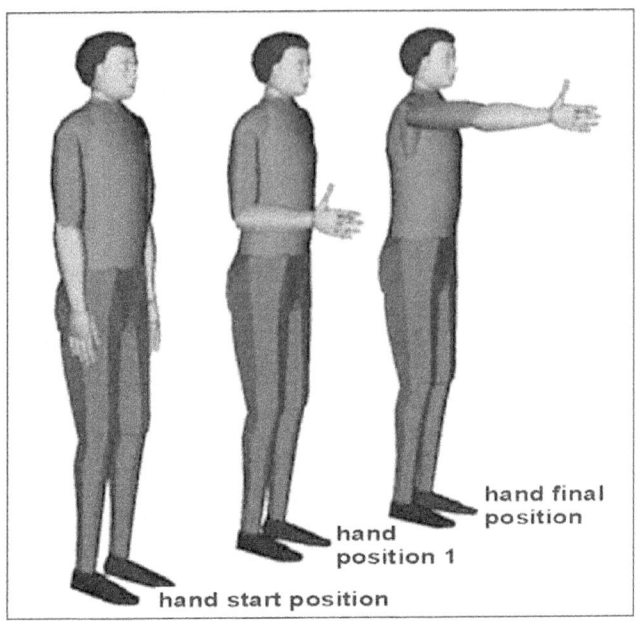

Figure 5.–Start-to-finish hand motions.

observed that all standard deviations were close to the assumed variance of motion used in preliminary testing of the random number generation in the code for virtual human motions. For example, motion envelop test results for an operator in a standing work posture revealed that 72% of the motions generated for the human subject's head, left hand, and left knee and 90% of the subject's starting positions were within 3 cm of the accepted standard deviation criteria. Manufacturer specifications of the motion tracking sensors used for tracking and capturing human movements had 3-cm accuracy; therefore, researchers used this value as a criterion for accepting standard deviations. These standard deviations provided the random seed numbers (N_1 in equation 3) for the random human motion code. The random seed numbers were placed in lookup tables within the program so they could be conveniently reused by motion commands.

LISP and JCL random motion code of the cube model and preliminary virtual human models evolved into the roof bolter model. Experimenting with basic motions in preliminary models helped researchers devise a random motion technique that worked with xyz values from basic motion parameters. Both values were the building blocks used in motion commands that comprised the core of the random motion code. Significant modifications expanded the code to create random figure movement, random motion goals for the arms and pelvis, and random motion of events reflecting operator work behavior. In the case of the roof bolter model, randomness was applied only to critical movements of the human figure, pelvis, and arm. Researchers were interested in contacts occurring when the machine appendage was moving, so this limited when and which body parts were critical. Code samples for each of these body parts follow with no additional description. The intent is simply to show the use of the random function in the motion command.

```
(jcl 'create_figure_motion "operator" "1.0sec" "1.5sec" "constant"
(jmult (jxyz -91.4 +4.63 (+ (- 97.77 spRZ) (random seedspRZ)))
(jtrans (+ (- 21.79 spX) (random seedspX)) +10.92 (+ (- 89.12 spZ) (random seedspZ)))))

(jcl 'create_pelvis_motion "operator" "7.0sec" "7.9sec" "constant"
(jmult (jxyz -90.00 (+ (- -2.27 H2) (random seedH2)) 96.18)
(jtrans 16.71 35.94 81.28)) "5.00" "decay")

(jcl 'create_arm_motion "operator" "7.0sec" "7.17sec" "constant" "left" "palm"
(jmult (jxyz +120.35 +75.97 +142.84)
(jtrans (+ (- -17.17 L2) (random seedL2)) (+ (- 92.25 L2) (random seedL2)) (+ (- 14.88 L2) (random seedL2))))
"no" "2.0" "constant")
```

The above sample "create_figure_motion" command used both xyz-orientation angles and xyz-positional coordinates. The "create_pelvis_motion" command used only the xyz-orientation angles, and the "create_arm_motion" command used only the xyz-positional coordinates. The manipulation value produced by the random motion technique for human motions is represented in equation 3 by modifying equation 1. Again using positional "X" as an example:

$$\{positional \ldots X\} \rightarrow (\pm N_0 \, N_1) - [\,a\ random\ number\ from\ zero\ to\ N\,], \quad (3)$$

where N_0 is the original X value from a basic motion scenario,

N_1 is the standard deviation value from experiments on motion envelopes, and

N is the N_1 value multiplied by two.

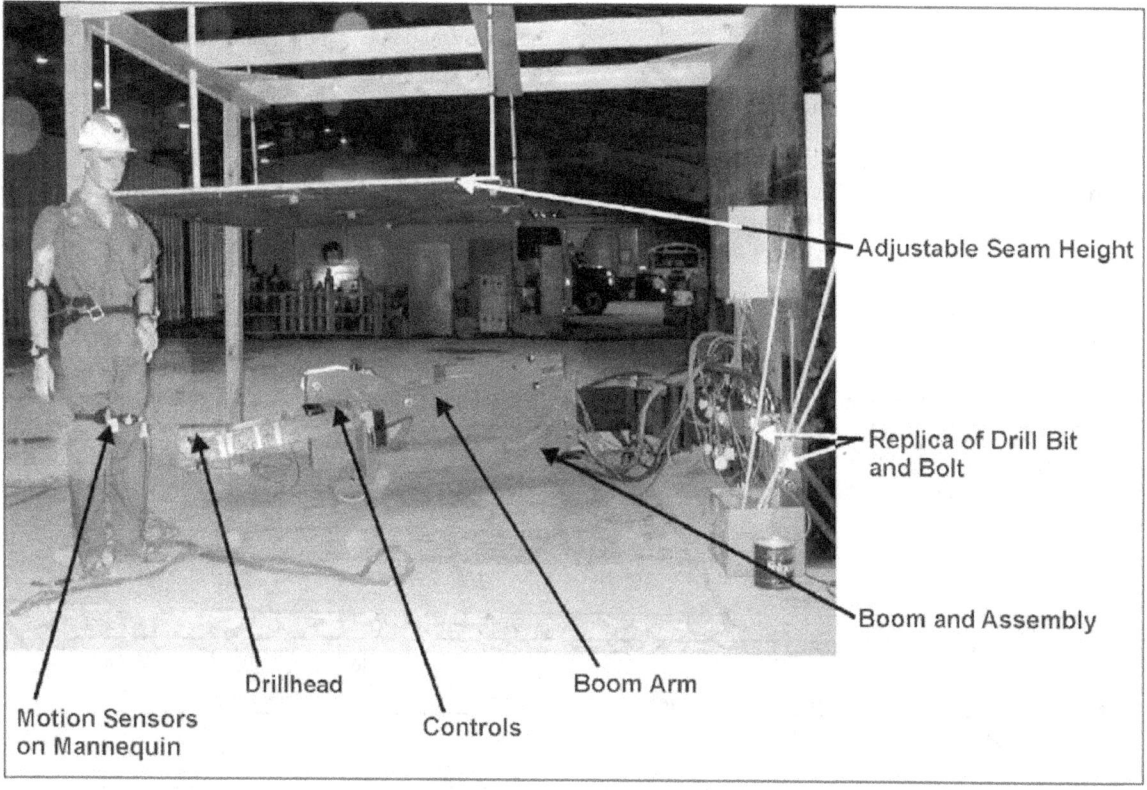

Figure 6.–Full-scale roof bolter assembly setup for data collection. A mannequin illustrates motion sensor locations on human subject.

The xyz values were obtained from the basic motions, and seed variables (special names such as spRZ, seedspRZ, spX, seedspX, spZ, seedspZ, H2, L2, seedH2, and seedL2) were standard deviation values from results obtained from the motion analysis by Bartels et al. [2001]. Lookup tables imbedded in the code defined the seed variables.

When human and machine models interact, the initial positioning of the operator relative to the machine determines the range of motion required to complete the task. The initial position limits the distance an operator can reach and establishes the starting point for the required path of operator movement. The initial position of the operator was also random.

In brief, the basic motions and standard derivation data provided random parameters for the operator's movement in the random motion code. When the code is executed in a model, it generates—

- Realistic starting positions for the operator (virtual human);
- Realistic starting positions in a simulated bolting sequence;
- A valid range of variation in movements; and
- Initial position for both the body and body parts.

RANDOM MOTION RULES

Researchers developed several rules for the random motion technique that, when applied, will maintain directional integrity of a virtual human's basic motions. The rules were developed from watching operators' basic motions during roof bolter simulations. Through these observations, it was easy to identify the principal direction of a basic motion, which helped define the motion type. Also, the observations helped to recognize the motion's direction and how a random change should affect the basic motion.

Before applying the rules, the resulting direction of all the basic motions of the virtual human operating the machine must be well defined and understood. The rules help researchers decide where to apply random motion in a motion command as to each of the motion's manipulation values for xyz-orientation angles and xyz-positional coordinates (table 1). Other possible combinations were omitted from the table because they were not needed in the roof-bolting scenario.

By applying the rules to the previous motion command samples, the "create_figure_motion" command reflects a starting position motion that contains a random "z" orientation angle and a random "x" and "z" for positional coordinates. The "create_pelvis_motion" command is an example for leaning forward and backward, and it contains only a "y" orientation angle. The "create_arm_motion" command is a point-to-point motion containing only random positional coordinates "x", "y", and "z".

RANDOM BEHAVIOR MOTION

A behavior motion is a series of human motions that mimics a specific action. Studies on worker job performance and on machinery and the work environment have identified miners' hazardous exposures while bolting [Klishis et al. 1993a,b]. The highest percentage of unsafe acts was found in two bolter tasks: drilling the hole and installing a bolt [Klishis et al. 1993a]. Using this information, researchers identified specific risky work behaviors for the drilling operation and bolt installation (table 2). Researchers also identified a nonrisky behavior for both bolter tasks. This behavior happens when none of the other risky behaviors occur. Researchers were interested in behaviors occurring only when the machine appendage had movement (figures 7-9). Subsequently, other risky behaviors associated with drilling the hole or installing a bolt were not used.

Eight human motion scenarios were developed that reflected all behavioral combinations used in the roof bolter model. Four separate human motion scenarios were needed to cover each of the two bolter tasks: drilling the hole and installing the bolt. To mimic the work behaviors in each of the scenarios, researchers modified the basic motions and preliminary models. Again, special care was taken in not allowing contacts to occur. All motion commands in the eight motion scenarios were converted into random motions. Appendix A contains code samples for each of the human motion scenarios.

Table 1.—Random motion rules for a roof bolter machine

Motion's direction	Operator's motion type						
	Basic	Prominent vertical direction	Vertical with z direction	Vertical with x direction	Point to point	Leaning forward and backward	Starting position
Orientation angles.	x y z	x y z	x y z	x y z	x y z	x random y z	x y random z
Positional coordinates.	x y z	x random y z	x random y random z	random x random y z	random x random y random z	x y z	random x y random z

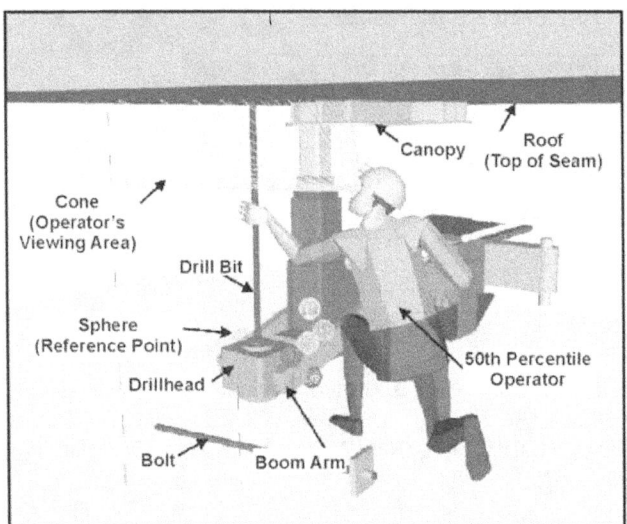

Figure 7.–Virtual operator's behavior hand on drill bit.

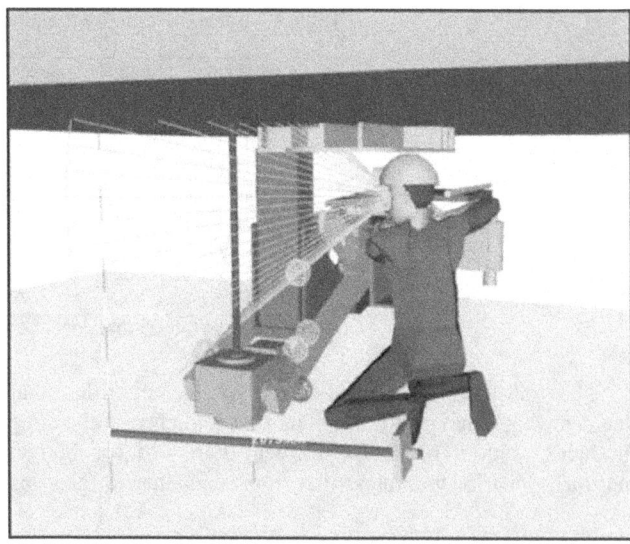

Figure 9.–Virtual operator's behavior hand off drill and boom arm.

Figure 8.–Virtual operator's behavior hand on boom arm.

A decision algorithm was needed for the random motion code to randomly select what behavior to use for a simulation execution. Numerical parameters used in the algorithm came from the percentage of operator actions that resulted in hazard exposure. These parameters were based on statistical observations of bolter operator actions associated with unsafe acts [Klishis et al. 1993b]. Specific unsafe acts and their percentages that could cause injuries or fatalities in a bolter's workspace include [Klishis et al. 1993b]:

- Drilling
  - Hand on the drill steel bit (HOSB - 18%)
  - Hand on the mast or boom arm (HOM - 59%)
- Bolt installation
  - Hand on the bolt or wrench (HOBW - 33%)
  - Hand on the mast or boom arm (HOMB - 49%)

Using this information, researchers formulated the algorithm that determines the behavior during the drilling operation and installation of the bolt (table 3).

Table 2.—Work behavior list for drilling the hole and installing a bolt

Operation	Code name	Behavior	Risky
Drill	BEHD1	Hand off the boom arm and hand off the drill steel bit	No.
	BEHD2	Hand on the drill steel bit	Yes.
	BEHD3	Hand on the boom arm	Yes.
	BEHD4	Hand on the boom arm and then hand on the drill steel bit	Yes.
Bolt	BEHB1	Hand off the boom arm and hand off the bolt or wrench	No.
	BEHB2	Hand on the bolt or wrench	Yes.
	BEHB3	Hand on the boom arm	Yes.
	BEHB4	Hand on the boom arm and then on the bolt or wrench	Yes.

Table 3.—Decision logic for the algorithm to determine behavior selection

Operation	Human motion scenario[1]	Mathematical relations symbolizing unsafe acts			
		HOSB	HOM	HOBW	HOMB
Drill	BEHD1	>18	>59	—	—
	BEHD2	≤18	>59	—	—
	BEHD3	>18	≤59	—	—
	BEHD4	≤18	≤59	—	—
Bolt	BEHB1	—	—	>33	>49
	BEHB2	—	—	≤33	>49
	BEHB3	—	—	>33	≤49
	BEHB4	—	—	≤33	≤49

[1] Code names from table 2 that describe the virtual operator's risky work behavior.

The following code is a sample of the algorithm that decides what behavior to use. The first sequence of the code produces random numbers between 0 to 100 for human motion scenario variables that represent unsafe acts as discussed above and reflected in table 3: HOSB, HOM, HOBW, and HOMB. A value is recalculated for HOSB, HOM, HOBW, and HOMB each time a new simulation is executed.

```
;; DRILLING - RANDOM NUMBER GENERATION FOR BEHAVIOR
(def hosb  (+ 0 (random 100.0))); hands on steel 18%
(def hom   (+ 0 (random 100.0))); hands on mast  59%
;; BOLTING - RANDOM NUMBER GENERATION FOR BEHAVIOR
(def hobw  (+ 0 (random 100.0))); hands on bolt/wrench 33%
(def homb  (+ 0 (random 100.0))); hands on mast installing bolt  49% ....
```

The second sequence of the code decides according to the values of HOSB, HOM, HOBW, and HOMB which virtual human motion scenario is used. The following is a sample of the programming code for each decision case. (*NOTE:* The specific motion commands are not listed.)

```
;;DRILLING NORM  MOTIONS (no risk)
(when (> hosb 18)(when (> hom 59) ....

;; DRILLING HOSB - HAND ON STEEL BIT MOTIONS
(when (<= hosb 18)(when (> hom 59) ....

;; DRILLING HOM - HAND ON BOOM ARM MOTIONS
(when (> hosb 18)(when (<= hom 59) ....

;; DRILLING BOTH, HAND ON STEEL BIT AND BOOM ARM MOTIONS
(when (<= hosb 18)(when (<= hom 59) ....

;; BOLTING NORM  MOTIONS (no risk)
(when (> hobw 33)(when (> homb 49) ....

;; BOLTING HOBW - HAND ON BOLT MOTIONS
(when (<= hobw 33)(when (> homb 49) ....

;; BOLTING HOM - HAND ON BOOM ARM MOTIONS WHEN INSTALLING THE BOLT
(when (> hobw 33)(when (<= homb 49) ....

;; BOLTING BOTH, HAND ON BOLT AND BOOM ARM MOTIONS
(when (<= how 33)(when (<= hosb 49) ....
```

The decision algorithm generates 1 of the possible 16 scenario combinations. Thus, random virtual human motion combinations give a realistic representation of the operator's motions and risky behaviors during drilling and bolting tasks.

VIRTUAL HUMAN ATTRIBUTES: VIEWING AREA

The operator's chance of avoiding a contact was also a concern to ensure that data reflecting a near-miss would not be considered a contact. This required knowledge of when the virtual operator sees the moving boom arm and the operator's reaction time to get out of the way of the boom arm. Therefore, to filter the database obtained from simulations for actual contacts, it was necessary to track the operator's field of view and determine from this information when the boom arm and drillhead are in and out of the operator's view.

Researchers originally used a viewing area that was a cone model with an oval directrix as defined by Humantech [1996] to experiment with the virtual human's vision tracking capabilities. MSHA's minimum lighting requirements mandate illumination levels of 0.06 fL for acceptable viewing in reduced lighting conditions found in underground mines. The viewing area was modified from test results from Bartels et al. [2001] on human subjects that determined the optimal viewing area and accurate field of vision for the virtual human. The results of analysis were averaged for the subjects and a vision area for the unique lighting conditions of underground mining environments. The vision area accounted for the use of a cap lamp and the reduction of viewing area by the use of a standard hard hat. The results of the tests in 0.06-fL lighting with a cap lamp and hard hat were the most significant in terms of input to the virtual human model. The most significant reduction in a subject's vision cone appeared to be a result of the reduction of the viewing area caused by the hard hat. The rods of the eye, which become more active in low light and allow night vision, were also the most sensitive to movement. The response of the eye rods was only slightly diminished. Figure 10 compares the directrix of the viewing area's cone of the original viewing area (simulation) to the final area (0.06 fL with a cap lamp on a hard hat). The normal light area is without the hard hat.

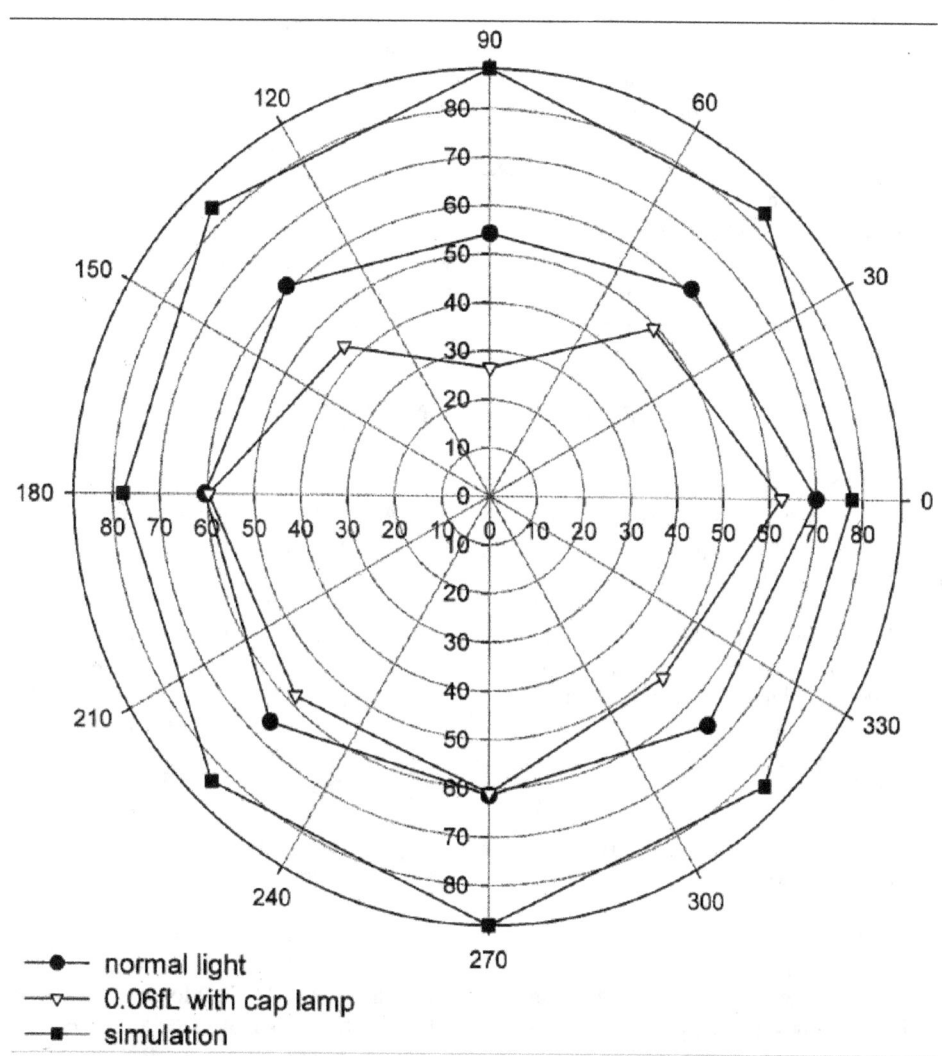

Figure 10.—Angular data of the original (normal light) and modified (0.06 fL with cap lamp) viewing areas for the virtual operator. Preliminary tests used the "simulation viewing" cone.

Test results from Bartels et al. [2001] helped reshape the viewing area. The new viewing area's directrix was reduced. The directrix became smaller and heart-shaped because the bill of the hard hat caused an obstruction (see graph in figure 10 of 0.06 fL with a cap lamp).

From this newly shaped viewing area, researchers developed a scheme that determined within the model when the boom arm and drillhead would be in and out of the virtual operator's view. The scheme used calculated distances from reference points placed on the viewing area and boom arm. Researchers were interested in contacts occurring when the machine appendage had movement, so this helped limit the area for tracking objects and enhanced the accuracy of the newly shaped viewing area. Given the location of the virtual operator from the boom arm during operation, the ideal length of the cone-shaped viewing area was 100 cm (39 in). The irregular shape of the directrix provided eight locations for reference points (figure 11). The boom arm reference point in the model was 23 cm (9 in) from the edge of the farthest boom support arm from the virtual operator. This location was ideal because it was the best position on the boom arm relative to the center of the field of view of the cone that would accommodate different operator work postures, such as kneeling and standing. The vertex of the cone-shaped viewing area was attached to the eyeballs of the virtual human. This allowed the viewing area to follow movements of the head and eyeballs when motion commands directed the virtual human to track or look at objects in the virtual environment.

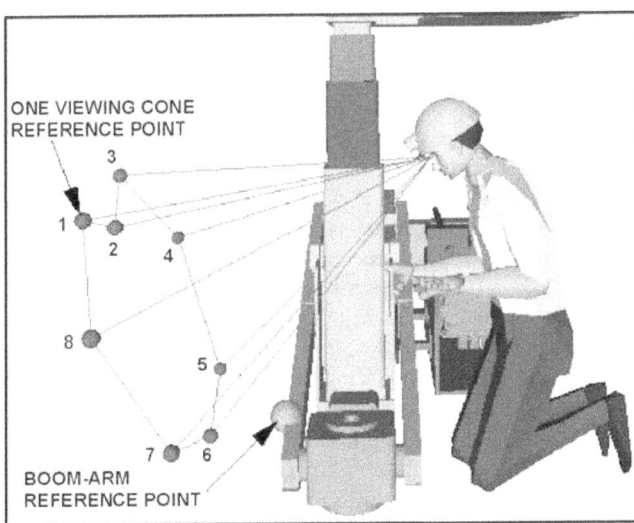

Figure 11.—Reference points on the viewing cone and boom arm.

In the program code, distance calculations between the viewing cone reference points and the boom arm reference were made and recorded every 0.03 sec. Experiments to validate the scheme showed that when one or more of the eight resulting distances were negative, the boom arm and drillhead were out of view of the virtual operator. Subsequently, all eight resulting distances must be positive for the scheme to indicate that the boom arm and drillhead are in view. This information is a crucial contribution to make a realistic representation of the operator's contact with the machine during drilling and bolting tasks.

CALCULATIONS AND OUTPUT

Jack's motion system is based on a notion of time. The actual movement of a motion system sequence, as a set of discrete frames, is generated by simulating time. The calculations and output section of the random code produce information every 0.03 sec. In 1 simulated second, there are 30 discrete simulated frames. For every simulated frame (0.03 sec), the code calculates and records to a file (table 4) the following information:

- Simulated time, in seconds;
- The operator's distance (cm) from the boom arm to help determine the virtual operator's location from the boom arm prior to performing tasks;
- The boom arm distance (cm) from a reference point on the floor level to help decide arm movement;
- Distance calculations (cm) between eight viewing area reference points and a reference point on the boom arm to help determine when the virtual operator sees the boom arm; and

- A number marking sequential contacts between virtual limbs and machine appendage for each simulated frame.

A simulated frame constitutes one line of data in the output file. Simulation information is printed in the first line of the output data file that defines the following:

- Machine control configuration;
- Seam height;
- Work posture;
- Boom arm speed;
- The operator's anthropometry; and
- The operator's behavior during the drilling and bolting tasks.

A code sample for the calculations and output section is found in appendix B. Additional code was developed and easily modified for output that calculates every simulated frame distance between body parts and reference points on the machine appendage. This information could be helpful for designing workstations that improve the operator's reach zone to depict the areas of maximum and comfortable reach for operators during machine operations.

Table 4.—Sample data output file

time	OPL	V1	V2	V3	V4	V5	V6	V7	V8	BAM	LPB	LPD	LAB	LAD	LLB	LLD	RLB	RLD	HDB
0.03	54.	-4.	38.	5.	63.	28.	58.	32.	20.	19.	0.	0.	0.	0.	0.	0.	0.	0.	0.
0.06	54.	-4.	38.	5.	63.	28.	58.	32.	20.	19.	0.	0.	0.	0.	0.	0.	0.	0.	0.
0.10	54.	-4.	38.	5.	63.	28.	58.	32.	20.	19.	0.	0.	0.	0.	0.	0.	0.	0.	0.
0.13	54.	-4.	38.	5.	63.	28.	58.	32.	20.	19.	0.	0.	0.	0.	0.	0.	0.	0.	0.
0.16	54.	-4.	38.	5.	63.	28.	58.	32.	20.	19.	0.	0.	0.	0.	0.	0.	0.	0.	0.
0.20	54.	-4.	38.	5.	63.	28.	58.	32.	20.	19.	0.	0.	0.	0.	0.	0.	0.	0.	0.
0.23	54.	-4.	38.	5.	63.	28.	58.	32.	20.	19.	0.	0.	0.	0.	0.	0.	0.	0.	0.

CONF=1 SEAM=2 POST=3 SPED=1 SUBJ=1 BEHD=1 BEHB=1

First line and columns of data file (coded)	Subsequent lines and columns in data file

CONF - machine control configuration
   1=piano key controls
SEAM - seam height
   1=144 cm (45 in); 2=152 cm (60 in); 3=182 cm (72 in)
POST - work posture
   1=right knee; 2=left knee; 3=both knees; 4=standing
SPED - boom arm speed
   1=18 cm/sec (7 in/sec); 2=25 cm/sec (10 in/sec)
   3=41 cm/sec (16 in/sec); 4=56 cm/sec (22 in/sec)
SUBJ - operator's anthropometry
   1=25th; 2=55th; 3=92nd
BEHD - operator's behavior during the drilling task
   1=none; 2=hand on drill
   3=hand on boom; 4=hand on both
BEHB - operator's behavior during the bolting task
   1=none; 2=hand on bolt
   3=hand on boom; 4=hand on both
	time - simulated time, sec
	OPL - operator's distance from the boom arm, cm
	V1 through V8 - reference points on the vision cone are used to evaluated if the boom arm is seen by the operator
	BAM - to determine boom arm movement, a distance is measured between a floor reference point and boom arm reference, cm
	LPH through HDB - a numerical marking that indicates if a contact occurred between an operator limb and machine appendage. "1" means contact; "0" means no contact.
	LPB / LPD = left palm with boom / with drillhead
	LAB / LAD = left forearm with boom / with drillhead
	LLB / LLD = left leg with boom / with drillhead
	RLB / RLD = right leg with boom / with drillhead
	HDB = head with boom arm

CONCLUSIONS

Researchers developed a program code that combines discrete-event and continuous simulation with built-in random manipulation values (xyz values) for a model before it is transformed into a motion path by Jack software's motion system. The manipulation values for xyz-orientation angles and xyz-positional coordinates define the final posture position of the virtual human. Understanding Jack's human motion kinematics and Jack's motion system were essential to random motion and behavior development. The random motions gave the model the realistic representation of the operator's motions and risky work behaviors found during anthropotechnical system tasks such as drilling and bolting. The results of this study were also successfully applied to a study that determined the optimal appendage speed range of a roof bolter so that the operators can do their work safely.

Successfully testing random motion within Jack's motion system was an important step toward its use on virtual human motions. Testing resulted in a primary programming structure and building blocks for code expansion to help researchers develop random movement in objects and calculate and store information on their action. The basic motions and standard derivation data provide random parameters of the operator's movement for the random motion program code. When the code is executed in the model, it generates for the operator (virtual human) realistic starting positions for a simulated bolting sequence and a valid range of variation in movements and initial position for both body and body parts.

Researchers developed several rules for the random motion technique that must be applied to maintain directional integrity of a virtual human's basic motions. These rules help decide where to apply random motion in a motion command as to each of the motion's manipulation values for xyz-orientation angles and xyz-positional coordinates.

A behavior motion is a series of human motions that mimic a specific action. Researchers were interested in behaviors occurring only when the machine appendage was moving. Eight human motion scenarios were developed that reflected all behavioral combinations. Motion commands in the eight virtual human motion scenarios were converted to random motions. Specific unsafe acts that could cause injuries or fatalities in a bolter's workspace were, during drilling, *hand on the drill steel bit* and *hand on the boom arm*, and, during bolt installation, *hand on the bolt or wrench* and *hand on the boom arm*. These data helped researchers to formulate an algorithm that determines the behavior during the simulation of the drilling operation and installation of the bolt.

Contacts might be avoided if the operator sees the moving appendage in time. A heart-shaped viewing area was developed from test results on human subjects that determined the optimal viewing area and accurate field of vision for the virtual human working in an underground mine environment. From this viewing area, a scheme was developed to determine when the boom arm and drillhead were in and out of the virtual operator's field of view. Also, the viewing area followed movements of the head and eyeballs when motion commands directed the virtual human to track or look at objects. Vision information is a crucial element that contributes to the realistic representation of the virtual operator and helps to determine when a contact is made with the appendage.

In the case of a roof bolter model, the program code calculates and records to a file the following: simulated time, operator's distance from the boom arm, the boom arm distance from a reference point to help decide arm movement, and distance calculations between the eight viewing area reference points and boom arm reference. Also included for each simulated frame is a numerical marking that indicates a contact, if any, between a body part and machine appendage.

Researchers in this study believe that the use of this random motion technique in simulations, treated with advanced statistical procedures such as survival analysis, provides extremely useful tools to evaluate the hazards of tasks where it is not possible or practical to perform experiments with human subjects. Furthermore, analysis of machine simulations can provide information that may be helpful in making recommendations to reduce the likelihood that operators experience an injury due to contact with a moving appendage. Such information can be useful for machine manufacturers for enhancing equipment designs and operational procedures.

ACKNOWLEDGMENTS

The author would like to thank Raymond F. Helinski, electronics technician (now retired), NIOSH Pittsburgh Research Laboratory, for his dedication, patience, and suggestions while running simulations that tested the code for random virtual human motions and behaviors. The author also thanks Chris Ege of UGS PLM Solutions for his instructions, suggestions, and guidance that helped to understand Jack software and for his help with JCL and LISP programming.

REFERENCES

Ambrose DH [2000]. A simulation approach analyzing random motion events between a machine and its operator. Warrendale, PA Society of Automotive Engineers, Inc., technical paper 2000-01-2160.

Ambrose DH, Bartels JR, Kwitowski AJ, Gallagher S, McWilliams LJ, Battenhouse TR Jr., et al. [forthcoming]. Mining roof bolting machine safety a study of the drill boom vertical velocity. Pittsburgh, PA U.S. Department of Health and Human Services, Public Health Service, Centers for Disease Control and Prevention, National Institute for Occupational Safety and Health, Information Circular (IC).

Azuola F, Badler NI, Ho P, Huh S, Kokkevis E, Ting B [1996]. Jack validation study final report. Philadelphia, PA University of Pennsylvania, Center for Human Modeling and Simulation. Grant No. DAMD 17-94-J-4486-P50002.

Bartels JR, Ambrose DH, Wang RC [2001]. Verification and validation of roof bolter simulation models for studying events between a machine and its operator. Warrendale, PA Society of Automotive Engineers, Inc., technical paper 2001-01-2088.

Grosso MR, Quach RD, Otani E, Zhao J, Wei S, Ho P-H, et al. [1989]. Anthropometry for computer graphics human figures. Philadelphia, PA University of Pennsylvania, Department of Computer and Information Science, technical report MS-CIS-89-71.

Humantech, Inc. [1996]. Ergonomic design guidelines for engineers. Ann Arbor, MI Humantech, Inc.

Klishis MJ, Althous RC, Layne LA, Lies GM [1993a]. Coal mine injury analysis a model for reduction through training. Morgantown, WV West Virginia University, Mining Extension Service.

Klishis MJ, Althous RC, Layne LA, Lies GM [1993b]. A manual for improving safety in roof bolting. Morgantown, WV West Virginia University, Mining Extension Service.

NASA [1987]. Man-systems integration standards. Hanover, MD National Aeronautics and Space Administration, Center for Aero Space Information, technical standard NASA-STD-3000, vol. 1.

Pandya A [1992a]. Correction and prediction of dynamic human joint strength from lean body mass. Hanover, MD National Aeronautics and Space Administration, Center for Aero Space Information, NASA technical paper 3207.

Pandya A [1992b]. The validation of a human force model to predict dynamic forces resulting from multi-joint motions. Hanover, MD National Aeronautics and Space Administration, Center for Aero Space Information, NASA technical paper 3206.

Steele GL Jr. [1990]. Common LISP the language. Woburn, MA Digital Press.

U.S. Army [1988]. Anthropometric survey of U.S. Army personnel methods and summary statistics. Natick, MA U.S. Army, Natick technical report TR-89-044.

APPENDIX A.—CODE SAMPLES OF RANDOM VIRTUAL HUMAN MOTION SCENARIOS OF BEHAVIOR

```
;; DRILLING NORM MOTIONS (no risk)
;; hand_L1
  (jcl 'create_arm_motion "operator" "6.0sec" "7.0sec" "constant" "left" "palm"
  (jmult (jxyz +125.25 +73.29 +129.45)
  (jtrans -18.21 (+ (- +65.86 L1) (random seedL1)) +19.52)) "no" "2.0" "constant")
;; hand_L2
  (jcl 'create_arm_motion "operator" "7.0sec" "7.17sec" "constant" "left" "palm"
  (jmult (jxyz +120.35 +75.97 +142.84)
  (jtrans (+ (- -17.17 L2) (random seedL2)) (+ (- 92.25 L2) (random seedL2)) (+ (- 14.88 L2)   (random seedL2)))) "no" "2.0" "constant")
;; hand_L3
  (jcl 'create_arm_motion "operator" "7.17sec" "7.34sec" "constant" "left" "palm"
  (jmult (jxyz +116.46 +77.46 +148.29)
  (jtrans (+ (- -17.48 L2) (random seedL2)) (+ (- 92.13 L2) (random seedL2)) (+ (- 49.94 L2)   (random seedL2)))) "no" "2.0" "constant")
;; hand_L4
  (jcl 'create_arm_motion "operator" "7.3sec" "7.47sec" "constant" "left" "palm"
  (jmult (jxyz +103.89 +76.59 +157.86)
  (jtrans (+ (- -10.33 L2) (random seedL2)) (+ (- 52.97 L2) (random seedL2)) (+ (- 52.68 L2)   (random seedL2)))) "no" "2.0" "constant")
;; hand_L5
  (jcl 'create_arm_motion "operator" "7.50sec" "7.97sec" "constant" "left" "palm"
  (jmult (jxyz +111.12 +76.47 +158.93)
  (jtrans (+ (- -10.94 L2) (random seedL2)) (+ (- 52.57 L2) (random seedL2)) (+ (- 56.99 L2)   (random seedL2)))) "no" "2.0" "constant")
;; hand_L6
  (jcl 'create_arm_motion "operator" "15.90sec" "16.17sec" "constant" "left" "palm"
  (jmult (jxyz +94.36 +80.31 +169.08)
  (jtrans (+ (- -25.06 L3) (random seedL3)) (+ (- 73.04 L3) (random seedL3)) (+ (- 51.42 L3)   (random seedL3)))) "no" "2.0" "constant")
;; hand_L7
  (jcl 'create_arm_motion "operator" "16.10sec" "16.30sec" "constant" "left" "palm"
  (jmult (jxyz +115.66 +72.16 +136.92)
  (jtrans (+ (- -21.69 L3) (random seedL3)) (+ (- 66.19 L3) (random seedL3)) (+ (- 19.26 L3)   (random seedL3)))) "no" "2.0" "constant")
;; hand_L8
  (jcl 'create_arm_motion "operator" "16.40sec" "16.80sec" "constant" "left" "palm"
  (jmult (jxyz +77.57 +31.48 +146.07)
  (jtrans -16.27 (+ (- 75.83 L4) (random seedL4)) 25.80))
   "no" "2.0" "constant")
;; pelvis1
  (jcl 'create_pelvis_motion "operator" "7.0sec" "7.9sec" "constant"
  (jmult (jxyz -90.00 (+ (- -2.27 H2) (random seedH2)) 96.18)
  (jtrans 16.71 35.94 81.28)) "5.00" "decay")
;; pelvis2
  (jcl 'create_pelvis_motion "operator" "10.87sec" "11.14sec" "constant"
  (jmult (jxyz -87.93 (+ (- -2.27 H2) (random seedH2)) 96.18)
  (jtrans 16.71 35.94 81.28)) "5.00" "decay")
;; pelvis3
  (jcl 'create_pelvis_motion "operator" "11.10sec" "11.37sec" "constant"
  (jmult (jxyz -90.00 (+ (- -0.00 H2) (random seedH2)) 96.18)
  (jtrans 16.71 35.94 81.28)) "5.00" "decay")
```

;; pelvis4
 (jcl 'create_pelvis_motion "operator" "28.20sec" "28.57sec" "constant"
 (jmult (jxyz -98.69 (+ (- 2.27 H5) (random seedH5)) 96.18)
 (jtrans 16.71 35.94 81.28)) "5.00" "decay")
;; pelvis5
(jcl 'create_pelvis_motion "operator" "28.70sec" "29.10sec" "constant"
 (jmult (jxyz -100.64 (+ (- 0.0 H5) (random seedH5)) 96.18)
 (jtrans 16.71 35.94 81.28)) "5.00" "decay")

;; DRILLING HOSB - HAND ON STEEL BIT MOTIONS
;; hand_L1
 (jcl 'create_arm_motion "operator" "6.0sec" "7.0sec" "constant" "left" "palm"
 (jmult (jxyz +125.25 +73.29 +129.45)
 (jtrans -18.21 (+ (- +65.86 L1) (random seedL1)) +19.52)) "no" "2.0" "constant")
;; hand_L2
 (jcl 'create_arm_motion "operator" "7.0sec" "7.90sec" "constant" "left" "palm"
 (jmult (jxyz +114.34 +61.20 +167.62)
 (jtrans (+ (- -17.74 L2) (random seedL2)) (+ (- 83.25 L2) (random seedL2)) (+ (- 15.80 L2)   (random seedL2)))) "no" "2.0" "constant")
;; hand_L3
 (jcl 'create_arm_motion "operator" "7.90sec" "8.50sec" "constant" "left" "palm"
 (jmult (jxyz +95.87 14.73 169.76)
 (jtrans (+ (- -16.03 L2) (random seedL2)) (+ (- 68.19 L2) (random seedL2)) (+ (- 40.68 L2)   (random seedL2)))) "no" "2.0" "constant")
;; hand_L4
 (jcl 'create_arm_motion "operator" "8.50sec" "9.0sec" "constant" "left" "palm"
 (jmult (jxyz +90.48 8.78 170.20)
 (jtrans (+ (- -15.49 L2) (random seedL2)) (+ (- 66.85 L2) (random seedL2)) (+ (- 54.81 L2)   (random seedL2)))) "no" "2.0" "constant")
;; hand_L5
 (jcl 'create_arm_motion "operator" "9.0sec" "9.50sec" "constant" "left" "palm"
 (jmult (jxyz +81.59 21.00 169.42)
 (jtrans (+ (- -0.75 L2) (random seedL2)) (+ (- 37.30 L2) (random seedL2)) (+ (- 62.02 L2)   (random seedL2)))) "no" "2.0" "constant")
;; hand_L6
 (jcl 'create_arm_motion "operator" "15.0sec" "15.90sec" "constant" "left" "palm"
 (jmult (jxyz +110.71 55.55 164.22)
 (jtrans (+ (- -12.71 L3) (random seedL3)) (+ (- 68.16 L3) (random seedL3)) (+ (- 23.34 L3)   (random seedL3)))) "no" "2.0" "constant")
;; hand_L7
 (jcl 'create_arm_motion "operator" "15.90sec" "16.17sec" "constant" "left" "palm"
 (jmult (jxyz 127.44 61.62 147.41)
 (jtrans -17.27 (+ (- 68.21 L3) (random seedL3)) +16.92)) "no" "2.0" "constant")
;; hand_L8
 (jcl 'create_arm_motion "operator" "16.17sec" "16.33sec" "constant" "left" "palm"
 (jmult (jxyz 133.12 64.21 142.59)
 (jtrans (+ (- -22.03 L3) (random seedL3)) (+ (- 76.08 L3) (random seedL3)) (+ (- 17.32 L3)   (random seedL3)))) "no" "2.0" "constant")
;; hand_L9
 (jcl 'create_arm_motion "operator" "16.40sec" "16.77sec" "constant" "left" "palm"
 (jmult (jxyz 77.57 31.48 146.07)
 (jtrans (+ (- -16.27 L3) (random seedL3)) (+ (- 75.83 L3) (random seedL3)) (+ (- 25.80 L3)   (random seedL3)))) "no" "2.0" "constant")
;; pelvis1
 (jcl 'create_pelvis_motion "operator" "7.9sec" "8.8sec" "constant"
 (jmult (jxyz -90.00 (+ (- -2.27 H2) (random seedH2)) 96.18)
 (jtrans 16.71 35.94 81.28)) "5.00" "decay")

```
;; pelvis2
  (jcl 'create_pelvis_motion "operator" "10.87sec" "11.14sec" "constant"
  (jmult (jxyz -87.93 (+ (- -2.27 H2) (random seedH2)) 96.18)
  (jtrans 16.71 35.94 81.28)) "5.00" "decay")
;; pelvis3
  (jcl 'create_pelvis_motion "operator" "11.10sec" "11.37sec" "constant"
  (jmult (jxyz -90.00 (+ (- -0.00 H2) (random seedH2)) 96.18)
  (jtrans 16.71 35.94 81.28)) "5.00" "decay")
;; pelvis4
  (jcl 'create_pelvis_motion "operator" "28.20sec" "28.57sec" "constant"
  (jmult (jxyz -98.69 (+ (- 2.27 H5) (random seedH5)) 96.18)
  (jtrans 16.71 35.94 81.28)) "5.00" "decay")
;; pelvis5
  (jcl 'create_pelvis_motion "operator" "28.70sec" "29.10sec" "constant"
  (jmult (jxyz -90.00 (+ (- 0.0 H5) (random seedH5)) 96.18)
  (jtrans 22.79 10.92 90.12)) "5.00" "decay")

;; DRILLING HOM - HAND ON BOOM-ARM MOTIONS
;; hand_L1
  (jcl 'create_arm_motion "operator" "6.0sec" "7.0sec" "constant" "left" "palm"
  (jmult (jxyz +125.25 +73.29 +129.45)
  (jtrans -18.21 (+ (- +65.86 L1) (random seedL1)) +19.52)) "no" "2.0" "constant")
;; hand_L2
  (jcl 'create_arm_motion "operator" "7.0sec" "7.17sec" "constant" "left" "palm"
  (jmult (jxyz +120.35 75.97 142.84)
  (jtrans (+ (- -17.74 L2) (random seedL2)) (+ (- 92.25 L2) (random seedL2)) (+ (- 14.88 L2)   (random seedL2)))) "no" "2.0" "constant")
;; hand_L3
  (jcl 'create_arm_motion "operator" "7.17sec" "7.27sec" "constant" "left" "palm"
  (jmult (jxyz 95.11 22.75 168.45)
  (jtrans (+ (- -22.75 L2) (random seedL2)) (+ (- 34.96 L2) (random seedL2)) (+ (- 34.85 L2)   (random seedL2)))) "no" "2.0" "constant")
;; hand_L4
  (jcl 'create_arm_motion "operator" "7.90sec" "8.50sec" "constant" "left" "palm"
  (jmult (jxyz 93.69 9.22 168.61)
  (jtrans (+ (- -15.57 L2) (random seedL2)) (+ (- 55.93 L2) (random seedL2)) (+ (- 40.76 L2)   (random seedL2)))) "no" "2.0" "constant")
;; hand_L5
  (jcl 'create_arm_motion "operator" "8.50sec" "9.0sec" "constant" "left" "palm"
  (jmult (jxyz 96.01 6.75 167.58)
  (jtrans (+ (- -15.55 L2) (random seedL2)) (+ (- 66.65 L2) (random seedL2)) (+ (- 40.89 L2)   (random seedL2)))) "no" "2.0" "constant")
;; hand_L6
  (jcl 'create_arm_motion "operator" "9.0sec" "9.50sec" "constant" "left" "palm"
  (jmult (jxyz 95.24 7.00 168.55)
  (jtrans (+ (- -15.06 L2) (random seedL2)) (+ (- 75.04 L2) (random seedL2)) (+ (- 40.80 L2)   (random seedL2)))) "no" "2.0" "constant")
;; hand_L7
  (jcl 'create_arm_motion "operator" "9.50sec" "10.0sec" "constant" "left" "palm"
  (jmult (jxyz 96.95 6.80 166.95)
  (jtrans (+ (- -15.44 L2) (random seedL2)) (+ (- 82.28 L2) (random seedL2)) (+ (- 40.69 L2)   (random seedL2)))) "no" "2.0" "constant")
;; hand_L8
  (jcl 'create_arm_motion "operator" "10.0sec" "10.50sec" "constant" "left" "palm"
  (jmult (jxyz 94.67 9.83 166.69)
  (jtrans (+ (- -15.57 L2) (random seedL2)) (+ (- 91.70 L2) (random seedL2)) (+ (- 40.77 L2)   (random seedL2)))) "no" "2.0" "constant")
```

```
;; hand_L9
  (jcl 'create_arm_motion "operator" "10.50sec" "10.87sec" "constant" "left" "palm"
  (jmult (jxyz 102.74 6.92 166.60)
  (jtrans (+ (- -15.86 L2) (random seedL2)) (+ (- 95.69 L2) (random seedL2)) (+ (- 48.18 L2)   (random seedL2))))  "no" "2.0" "constant")
;; hand_L10
  (jcl 'create_arm_motion "operator" "12.80sec" "13.20sec" "constant" "left" "palm"
  (jmult (jxyz 99.66 6.48 168.95)
  (jtrans (+ (- 0.44 L3) (random seedL3)) (+ (- 92.30 L3) (random seedL3)) (+ (- 36.04 L3)   (random seedL3))))  "no" "2.0" "constant")
;; hand_L11
  (jcl 'create_arm_motion "operator" "13.20sec" "14.0sec" "constant" "left" "palm"
  (jmult (jxyz 91.80 6.24 170.28)
  (jtrans (+ (- 0.69 L3) (random seedL3)) (+ (- 78.32 L3) (random seedL3)) (+ (- 36.35 L3)   (random seedL3))))  "no" "2.0" "constant")
;; hand_L12
  (jcl 'create_arm_motion "operator" "14.0sec" "15.0sec" "constant" "left" "palm"
  (jmult (jxyz +97.33 9.31 166.84)
  (jtrans (+ (- -15.74 L3) (random seedL3)) (+ (- 62.66 L3) (random seedL3)) (+ (- 37.75 L3)   (random seedL3))))  "no" "2.0" "constant")
;; hand_L13
  (jcl 'create_arm_motion "operator" "15.0sec" "15.90sec" "constant" "left" "palm"
  (jmult (jxyz 100.55 9.74 167.08)
  (jtrans (+ (- -15.82 L3) (random seedL3)) (+ (- 44.92 L3) (random seedL3)) (+ (- 37.66 L3)   (random seedL3))))  "no" "2.0" "constant")
;; hand_L14
  (jcl 'create_arm_motion "operator" "15.90sec" "16.17sec" "constant" "left" "palm"
  (jmult (jxyz 94.36 80.31 169.08)
  (jtrans (+ (- -25.06 L3) (random seedL3)) (+ (- 73.04 L3) (random seedL3)) (+ (- 51.42 L3)   (random seedL3))))  "no" "2.0" "constant")
;; hand_L15
  (jcl 'create_arm_motion "operator" "16.17sec" "16.37sec" "constant" "left" "palm"
  (jmult (jxyz 115.66 72.16 136.92)
  (jtrans -21.69 (+ (- 66.19 L3) (random seedL3)) +19.26)) "no" "2.0" "constant")
;; hand_L16
  (jcl 'create_arm_motion "operator" "16.40sec" "16.80sec" "constant" "left" "palm"
  (jmult (jxyz 77.57 31.48 146.07)
  (jtrans (+ (- -16.27 L3) (random seedL3)) (+ (- 75.83 L3) (random seedL3)) (+ (- 25.80 L3)   (random seedL3))))  "no" "2.0" "constant")
;; pelvis1
   (jcl 'create_pelvis_motion "operator" "7.0sec" "7.9sec" "constant"
   (jmult (jxyz -90.00 (+ (- -2.27 H2) (random seedH2)) 96.18)
   (jtrans 16.71 35.94 81.28)) "5.00" "decay")
;; pelvis2
   (jcl 'create_pelvis_motion "operator" "10.87sec" "11.14sec" "constant"
   (jmult (jxyz -87.93 (+ (- -2.27 H2) (random seedH2)) 96.18)
   (jtrans 16.71 35.94 81.28)) "5.00" "decay")
;; pelvis3
   (jcl 'create_pelvis_motion "operator" "11.10sec" "11.37sec" "constant"
   (jmult (jxyz -90.00 (+ (- -0.00 H2) (random seedH2)) 96.18)
   (jtrans 16.71 35.94 81.28)) "5.00" "decay")
;; pelvis4
   (jcl 'create_pelvis_motion "operator" "28.20sec" "28.57sec" "constant
   (jmult (jxyz -98.69 (+ (- 2.27 H5) (random seedH5)) 96.18)
   (jtrans 16.71 35.94 81.28)) "5.00" "decay")
;; pelvis5
   (jcl 'create_pelvis_motion "operator" "28.70sec" "29.10sec" "constant
```

```
  (jmult (jxyz -90.00 (+ (- 0.0 H5) (random seedH5)) 96.18)
  (jtrans 22.79 10.92 90.12)) "5.00" "decay")
```

;; DRILLING BOTH, HOSBM HANDS ON STEEL BIT AND BOOM-ARM MOTIONS
;; hand_L1
```
  (jcl 'create_arm_motion "operator" "6.0sec" "7.0sec" "constant" "left" "palm"
  (jmult (jxyz +125.25 +73.29 +129.45)
  (jtrans -18.21 (+ (- +65.86 L1) (random seedL1)) +19.52)) "no" "2.0" "constant")
```
;; hand_L2
```
  (jcl 'create_arm_motion "operator" "7.0sec" "7.90sec" "constant" "left" "palm"
  (jmult (jxyz +143.08 29.57 152.41)
  (jtrans -14.19 (+ (- +79.53 L2) (random seedL2)) +37.77)) "no" "2.0" "constant")
```
;; hand_L3
```
  (jcl 'create_arm_motion "operator" "7.90sec" "8.50sec" "constant" "left" "palm"
  (jmult (jxyz 93.69 9.22 168.61)
  (jtrans (+ (- -15.57 L2) (random seedL2)) (+ (- 55.93 L2) (random seedL2)) (+ (- 40.76 L2)   (random seedL2))))"no" "2.0" "constant")
```
;; hand_L4
```
  (jcl 'create_arm_motion "operator" "8.50sec" "9.0sec" "constant" "left" "palm"
  (jmult (jxyz 96.01 6.75 167.58)
  (jtrans (+ (- -15.55 L2) (random seedL2)) (+ (- 66.65 L2) (random seedL2)) (+ (- 40.89 L2)   (random seedL2)))) "no" "2.0" "constant")
```
;; hand_L5
```
  (jcl 'create_arm_motion "operator" "9.0sec" "9.50sec" "constant" "left" "palm"
  (jmult (jxyz 95.24 7.00 168.55)
  (jtrans (+ (- -15.06 L2) (random seedL2)) (+ (- 75.04 L2) (random seedL2)) (+ (- 40.80 L2)   (random seedL2)))) "no" "2.0" "constant")
```
;; hand_L6
```
  (jcl 'create_arm_motion "operator" "9.50sec" "10.0sec" "constant" "left" "palm"
  (jmult (jxyz 96.95 6.80 166.95)
  (jtrans (+ (- -15.44 L2) (random seedL2)) (+ (- 82.28 L2) (random seedL2)) (+ (- 40.69 L2)   (random seedL2)))) "no" "2.0" "constant")
```
;; hand_L7
```
  (jcl 'create_arm_motion "operator" "10.0sec" "10.50sec" "constant" "left" "palm"
  (jmult (jxyz 94.67 9.83 166.69)
  (jtrans (+ (- -15.57 L2) (random seedL2)) (+ (- 91.70 L2) (random seedL2)) (+ (- 40.77 L2)   (random seedL2)))) "no" "2.0" "constant")
```
;; hand_L8
```
  (jcl 'create_arm_motion "operator" "10.50sec" "10.87sec" "constant" "left" "palm"
  (jmult (jxyz 102.74 6.92 166.60)
  (jtrans (+ (- -15.86 L2) (random seedL2)) (+ (- 95.69 L2) (random seedL2)) (+ (- 48.18 L2)   (random seedL2)))) "no" "2.0" "constant")
```
;; hand_L9
```
  (jcl 'create_arm_motion "operator" "12.80sec" "13.20sec" "constant" "left" "palm"
  (jmult (jxyz 99.66 6.48 168.95)
  (jtrans (+ (- 0.44 L3) (random seedL3)) (+ (- 92.30 L3) (random seedL3)) (+ (- 36.04 L3)   (random seedL3)))) "no" "2.0" "constant")
```
;; hand_L10
```
  (jcl 'create_arm_motion "operator" "13.20sec" "14.0sec" "constant" "left" "palm"
  (jmult (jxyz 98.62 5.11 174.84)
  (jtrans (+ (- 0.24 L3) (random seedL3)) (+ (- 79.14 L3) (random seedL3)) (+ (- 35.72 L3)   (random seedL3)))) "no" "2.0" "constant")
```
;; hand_L11
```
  (jcl 'create_arm_motion "operator" "14.0sec" "15.0sec" "constant" "left" "palm"
  (jmult (jxyz +97.33 9.31 166.84)
  (jtrans (+ (- -15.74 L3) (random seedL3)) (+ (- 62.66 L3) (random seedL3)) (+ (- 37.75 L3)   (random seedL3)))) "no" "2.0" "constant")
```

```
;; hand_L12
  (jcl 'create_arm_motion "operator" "15.0sec" "15.90sec" "constant" "left" "palm"
  (jmult (jxyz 110.71 55.55 164.22)
  (jtrans -12.71 (+ (- +68.16 L3) (random seedL3)) +23.34)) "no" "2.0" "constant")
;; hand_L13
  (jcl 'create_arm_motion "operator" "15.90sec" "16.17sec" "constant" "left" "palm"
  (jmult (jxyz 127.96 63.62 144.90)
  (jtrans -21.83 (+ (- +68.27 L3) (random seedL3)) +17.00)) "no" "2.0" "constant")
;; hand_L14
  (jcl 'create_arm_motion "operator" "16.17sec" "16.37sec" "constant" "left" "palm"
  (jmult (jxyz 133.12 64.21 142.59)
  (jtrans -22.03 (+ (- +76.08 L3) (random seedL3)) +17.32)) "no" "2.0" "constant")
;; hand_L15
  (jcl 'create_arm_motion "operator" "16.40sec" "16.80sec" "constant" "left" "palm"
  (jmult (jxyz 77.57 31.48 146.07)
  (jtrans (+ (- -16.27 L3) (random seedL3)) (+ (- 75.83 L3) (random seedL3)) (+ (- 25.80 L3)   (random seedL3)))) "no" "2.0" "constant")
;; pelvis1
  (jcl 'create_pelvis_motion "operator" "7.9sec" "8.8sec" "constant"
  (jmult (jxyz -90.00 (+ (- -2.27 H2) (random seedH2)) 96.18)
  (jtrans 16.71 35.94 81.28)) "5.00" "decay")
;; pelvis2
  (jcl 'create_pelvis_motion "operator" "10.87sec" "11.14sec" "constant"
  (jmult (jxyz -87.93 (+ (- -2.27 H2) (random seedH2)) 96.18)
  (jtrans 16.71 35.94 81.28)) "5.00" "decay")
;; pelvis3
  (jcl 'create_pelvis_motion "operator" "11.10sec" "11.37sec" "constant"
  (jmult (jxyz -90.00 (+ (- -0.00 H2) (random seedH2)) 96.18)
  (jtrans 16.71 35.94 81.28)) "5.00" "decay")
;; pelvis4
  (jcl 'create_pelvis_motion "operator" "28.20sec" "28.57sec" "constant"
  (jmult (jxyz -98.69 (+ (- 2.27 H5) (random seedH5)) 96.18)
  (jtrans 16.71 35.94 81.28)) "5.00" "decay")
;; pelvis5
  (jcl 'create_pelvis_motion "operator" "28.70sec" "29.10sec" "constant"
  (jmult (jxyz -90.00 (+ (- 0.0 H5) (random seedH5)) 96.18)
  (jtrans 22.79 10.92 90.12)) "5.00" "decay")

;; BOLTING NORM MOTIONS (no risk)
;; hand_L9
  (jcl 'create_arm_motion "operator" "24.0sec" "24.20sec" "constant" "left" "palm"
  (jmult (jxyz +118.86 +79.68 +142.78)
  (jtrans -17.40 (+ (- 80.59 L4) (random seedL4)) +49.76)) "no" "2.0" "constant")
;; hand_L10
  (jcl 'create_arm_motion "operator" "24.2sec" "24.7sec" "constant" "left" "palm"
  (jmult (jxyz +130.02 +74.15 +132.64)
  (jtrans (+ (- -22.60 L5) (random seedL5)) (+ (- 53.93 L5) (random seedL5)) (+ (- 58.62 L5)   (random seedL5)))) "no" "2.0" "constant")

;; BOLTING HOBW - HAND ON BOLT MOTIONS
;; hand_L10
  (jcl 'create_arm_motion "operator" "24.0sec" "24.20sec" "constant" "left" "palm"
  (jmult (jxyz 97.26 59.59 162.16)
  (jtrans -14.47 (+ (- 64.01 L4) (random seedL4)) +12.97)) "no" "2.0" "constant")
;; hand_L11
  (jcl 'create_arm_motion "operator" "24.20sec" "25.0sec" "constant" "left" "palm"
  (jmult (jxyz 97.03 27.60 170.19)
```

(jtrans (+ (- -16.48 L5) (random seedL5)) (+ (- 72.89 L5) (random seedL5)) (+ (- 16.17 L5)   (random seedL5)))) "no" "2.0" "constant")
;; hand_L12
   (jcl 'create_arm_motion "operator" "25.0sec" "25.80sec" "constant" "left" "palm"
   (jmult (jxyz 121.23 40.60 177.74)
   (jtrans (+ (- -18.63 L5) (random seedL5)) (+ (- 83.35 L5) (random seedL5)) (+ (- 15.39 L5)   (random seedL5)))) "no" "2.0" "constant")
;; hand_L13
   (jcl 'create_arm_motion "operator" "28.50sec" "26.60sec" "constant" "left" "palm"
   (jmult (jxyz 136.78 47.95 162.95)
   (jtrans (+ (- -7.52 L5) (random seedL5)) (+ (- 78.79 L5) (random seedL5)) (+ (- 33.10 L5)   (random seedL5)))) "no" "2.0" "constant")
;; hand_L14
   (jcl 'create_arm_motion "operator" "26.60sec" "27.40sec" "constant" "left" "palm"
   (jmult (jxyz 105.50 38.04 168.90)
   (jtrans (+ (- -13.59 L5) (random seedL5)) (+ (- 69.55 L5) (random seedL5)) (+ (- 53.05 L5)   (random seedL5)))) "no" "2.0" "constant")
;; hand_L15
   (jcl 'create_arm_motion "operator" "27.40sec" "28.20sec" "constant" "left" "palm"
   (jmult (jxyz 46.22 38.22 -166.29)
   (jtrans (+ (- -4.03 L5) (random seedL5)) (+ (- 20.31 L5) (random seedL3)) (+ (- 51.81 L5)   (random seedL5)))) "no" "2.0" "constant")

;; BOLTING HOM - HAND ON BOOM-ARM MOTIONS WHEN INSTALLING THE BOLT
;; hand_L17
   (jcl 'create_arm_motion "operator" "24.0sec" "24.20sec" "constant" "left" "palm"
   (jmult (jxyz 75.05 27.68 173.15)
   (jtrans -16.03 (+ (- 37.03 L4) (random seedL4)) +36.98)) "no" "2.0" "constant")
;; hand_L18
   (jcl 'create_arm_motion "operator" "24.20sec" "25.0sec" "constant" "left" "palm"
   (jmult (jxyz 85.09 32.46 160.71)
   (jtrans (+ (- -21.44 L5) (random seedL5)) (+ (- 46.43 L5) (random seedL5)) (+ (- 36.61 L5)   (random seedL5)))) "no" "2.0" "constant")
;; hand_L19
   (jcl 'create_arm_motion "operator" "25.0sec" "25.80sec" "constant" "left" "palm"
   (jmult (jxyz 125.08 37.35 164.55)
   (jtrans (+ (- -22.65 L5) (random seedL5)) (+ (- 53.60 L5) (random seedL5)) (+ (- 37.60 L5)   (random seedL5)))) "no" "2.0" "constant")
;; hand_L20
   (jcl 'create_arm_motion "operator" "25.80sec" "26.60sec" "constant" "left" "palm"
   (jmult (jxyz 125.24 44.91 169.01)
   (jtrans (+ (- -8.67 L5) (random seedL5)) (+ (- 65.10 L5) (random seedL5)) (+ (- 37.47 L5)   (random seedL5)))) "no" "2.0" "constant")
;; hand_L21
   (jcl 'create_arm_motion "operator" "26.60sec" "27.40sec" "constant" "left" "palm"
   (jmult (jxyz 143.08 29.57 152.41)
   (jtrans (+ (- -14.19 L5) (random seedL5)) (+ (- 79.53 L5) (random seedL5)) (+ (- 37.77 L5)   (random seedL5)))) "no" "2.0" "constant")
;; hand_L22
   (jcl 'create_arm_motion "operator" "27.40sec" "28.20sec" "constant" "left" "palm"
   (jmult (jxyz +129.67 32.72 162.20)
   (jtrans (+ (- -5.80 L5) (random seedL5)) (+ (- 90.84 L5) (random seedL5)) (+ (- 37.46 L5)   (random seedL5)))) "no" "2.0" "constant")
;; hand_L23
   (jcl 'create_arm_motion "operator" "28.20sec" "28.70sec" "constant" "left" "palm"
   (jmult (jxyz 140.29 38.11 152.58)

     (jtrans (+ (- -9.28 L5) (random seedL5)) (+ (- 44.92 L5) (random seedL5)) (+ (- 50.05 L5)   (random seedL5)))) "no" "2.0" "constant")
;; hand_L24
   (jcl 'create_arm_motion "operator" "29.0sec" "29.50sec" "constant" "left" "palm"
   (jmult (jxyz 151.28 27.01 157.18)
     (jtrans (+ (- -16.23 L6) (random seedL6)) (+ (- 92.72 L6) (random seedL6)) (+ (- 40.78 L6)   (random seedL6)))) "no" "2.0" "constant")
;; hand_L25
   (jcl 'create_arm_motion "operator" "29.50sec" "30.50sec" "constant" "left" "palm"
   (jmult (jxyz 141.26 26.38 158.97)
     (jtrans (+ (- -14.94 L6) (random seedL6)) (+ (- 77.24 L6) (random seedL6)) (+ (- 40.66 L6)   (random seedL6)))) "no" "2.0" "constant")
;; hand_L26
   (jcl 'create_arm_motion "operator" "30.50sec" "31.50sec" "constant" "left" "palm"
   (jmult (jxyz 130.86 17.70 162.50)
     (jtrans (+ (- -10.23 L6) (random seedL6)) (+ (- 63.38 L6) (random seedL6)) (+ (- 37.85 L6)   (random seedL6)))) "no" "2.0" "constant")
;; hand_L27
   (jcl 'create_arm_motion "operator" "31.50sec" "32.50sec" "constant" "left" "palm"
   (jmult (jxyz 94.70 15.52 172.44)
     (jtrans (+ (- -5.32 L6) (random seedL6)) (+ (- 51.05 L6) (random seedL6)) (+ (- 34.41 L6)   (random seedL6)))) "no" "2.0" "constant")
;; hand_L28
   (jcl 'create_arm_motion "operator" "32.50sec" "33.50sec" "constant" "left" "palm"
   (jmult (jxyz 97.92 25.19 161.52)
     (jtrans (+ (- -22.11 L6) (random seedL6)) (+ (- 37.26 L6) (random seedL6)) (+ (- 38.80 L6)   (random seedL6)))) "no" "2.0" "constant")

;; BOLTING BOTH, HANDS ON BOLT AND BOOM-ARM MOTIONS
;; hand_L16
   (jcl 'create_arm_motion "operator" "24.0sec" "24.20sec" "constant" "left" "palm"
   (jmult (jxyz 97.26 59.59 162.16)
     (jtrans -14.47 (+ (- 64.01 L4) (random seedL4)) +12.97)) "no" "2.0" "constant")
;; hand_L17
   (jcl 'create_arm_motion "operator" "24.20sec" "25.0sec" "constant" "left" "palm"
   (jmult (jxyz 126.59 64.31 150.42)
     (jtrans (+ (- -16.22 L5) (random seedL5)) (+ (- 74.31 L5) (random seedL5)) (+ (- 13.17 L5)   (random seedL5)))) "no" "2.0" "constant")
;; hand_L18
   (jcl 'create_arm_motion "operator" "25.0sec" "25.80sec" "constant" "left" "palm"
   (jmult (jxyz 134.26 46.39 165.10)
     (jtrans (+ (- -19.62 L5) (random seedL5)) (+ (- 87.09 L5) (random seedL5)) (+ (- 18.21 L5)   (random seedL5)))) "no" "2.0" "constant")
;; hand_L19
   (jcl 'create_arm_motion "operator" "25.80sec" "26.60sec" "constant" "left" "palm"
   (jmult (jxyz 125.24 44.91 169.01)
     (jtrans (+ (- -8.67 L5) (random seedL5)) (+ (- 65.10 L5) (random seedL5)) (+ (- 37.47 L5)   (random seedL5)))) "no" "2.0" "constant")
;; hand_L20
   (jcl 'create_arm_motion "operator" "26.60sec" "27.40sec" "constant" "left" "palm"
   (jmult (jxyz 143.08 29.57 152.41)
     (jtrans (+ (- -14.19 L5) (random seedL5)) (+ (- 79.53 L5) (random seedL5)) (+ (- 37.77 L5)   (random seedL5)))) "no" "2.0" "constant")
;; hand_L21
   (jcl 'create_arm_motion "operator" "27.40sec" "28.20sec" "constant" "left" "palm"
   (jmult (jxyz +129.67 32.72 162.20)

(jtrans (+ (- -5.80 L5) (random seedL5)) (+ (- 90.84 L5) (random seedL5)) (+ (- 37.46 L5)   (random seedL5)))) "no" "2.0" "constant")
;; hand_L22
  (jcl 'create_arm_motion "operator" "28.20sec" "28.70sec" "constant" "left" "palm"
  (jmult (jxyz 140.29 38.11 152.58)
  (jtrans (+ (- -9.28 L5) (random seedL5)) (+ (- 44.92 L5) (random seedL5)) (+ (- 50.05 L5)   (random seedL5)))) "no" "2.0" "constant")
;; hand_L23
  (jcl 'create_arm_motion "operator" "29.0sec" "29.50sec" "constant" "left" "palm"
  (jmult (jxyz 151.28 27.01 157.18)
  (jtrans (+ (- -16.23 L6) (random seedL6)) (+ (- 92.72 L6) (random seedL6)) (+ (- 40.78 L6)   (random seedL6)))) "no" "2.0" "constant")
;; hand_L24
  (jcl 'create_arm_motion "operator" "29.50sec" "30.50sec" "constant" "left" "palm"
  (jmult (jxyz 141.26 26.38 158.97)
  (jtrans (+ (- -14.94 L6) (random seedL6)) (+ (- 77.24 L6) (random seedL6)) (+ (- 40.66 L6)   (random seedL6)))) "no" "2.0" "constant")
;; hand_L25
  (jcl 'create_arm_motion "operator" "30.50sec" "31.50sec" "constant" "left" "palm"
  (jmult (jxyz 130.86 17.70 162.50)
  (jtrans (+ (- -10.23 L6) (random seedL6)) (+ (- 63.38 L6) (random seedL6)) (+ (- 37.85 L6)   (random seedL6)))) "no" "2.0" "constant")
;; hand_L26
  (jcl 'create_arm_motion "operator" "31.50sec" "32.50sec" "constant" "left" "palm"
  (jmult (jxyz 94.70 15.52 172.44)
  (jtrans (+ (- -5.32 L6) (random seedL6)) (+ (- 51.05 L6) (random seedL6)) (+ (- 34.41 L6)   (random seedL6)))) "no" "2.0" "constant")
;; hand_L27
  (jcl 'create_arm_motion "operator" "32.50sec" "33.50sec" "constant" "left" "palm"
  (jmult (jxyz 97.92 25.19 161.52)
  (jtrans (+ (- -22.11 L6) (random seedL6)) (+ (- 37.26 L6) (random seedL6)) (+ (- 38.80 L6)   (random seedL6)))) "no" "2.0" "constant")

APPENDIX B.—CODE SAMPLES OF CALCULATIONS AND OUTPUT FOR THE RANDOM VIRTUAL HUMAN MOTION CODE

```
;;-------------------------------------------------------------------------
;; OUTPUT TO DATA FILE
;;-------------------------------------------------------------------------

(defun write ()
 (def trans_frames ())
 (def jnt_data ())
;;;; actual file name, i.e. ps4507R-0000.txt
 (def fig_name "ps4507R")
 (def channel_file
   (input-outfile "Enter filename:"
     (strcat fig_name "-0000") ".txt"))
 (def cf (open channel_file :direction :output))
;;
 (def counter 1)
    (format cf " CONF=~A " xconf)
    (format cf " SEAM=~A " xseam)
    (format cf " POST=~A " knee)
    (format cf " SPED=~A " xspeed)
    (format cf " SUBJ=~A " xsubj)
    (when (= xnorm 1) (format cf " BEHD=1 "))
    (when (= xhosb 1) (format cf " BEHD=2 "))
    (when (= xhom 1)  (format cf " BEHD=3 "))
    (when (= xboth 1) (format cf " BEHD=4 "))
    (when (= xnormb 1) (format cf " BEHB=1 \n"))
    (when (= xhobw 1)  (format cf " BEHB=2 \n"))
    (when (= xhomb 1)  (format cf " BEHB=3 \n"))
    (when (= xbothb 1) (format cf " BEHB=4 \n"))
;;;
    (format cf " time OPL V1 V2 V3 V4 V5 V6 V7 V8 BAM LPB LPD LAB LAD LLB LLD RLB RLD HDB \n")
;;;
;;Loop to print data
 (dotimes (count (+ frame_num 1))
;; sites
    (def posB1 (matrix-trans (aref B1 count)))
    (def posB2 (matrix-trans (aref B2 count)))
    (def posB3 (matrix-trans (aref B3 count)))
    (def posB4 (matrix-trans (aref B4 count)))
    (def posB5 (matrix-trans (aref B5 count)))
    (def posB6 (matrix-trans (aref B6 count)))
    (def posB7 (matrix-trans (aref B7 count)))
    (def posB8 (matrix-trans (aref B8 count)))
    (def posB9 (matrix-trans (aref B9 count)))
    (def posB10 (matrix-trans (aref B10 count)))
    (def posB11 (matrix-trans (aref B11 count)))
    (def posS1 (matrix-trans (aref subj_S1 count)))
    (def posS2 (matrix-trans (aref subj_S2 count)))
    (def posS3 (matrix-trans (aref subj_S3 count)))
    (def posS4 (matrix-trans (aref subj_S4 count)))
    (def posS5 (matrix-trans (aref subj_S5 count)))
    (def posS6 (matrix-trans (aref subj_S6 count)))
    (def posV1 (matrix-trans (aref subj_V1 count)))
```

```
      (def posV2 (matrix-trans (aref subj_V2 count)))
      (def posV3 (matrix-trans (aref subj_V3 count)))
      (def posV4 (matrix-trans (aref subj_V4 count)))
      (def posV5 (matrix-trans (aref subj_V5 count)))
      (def posV6 (matrix-trans (aref subj_V6 count)))
      (def posV7 (matrix-trans (aref subj_V7 count)))
      (def posV8 (matrix-trans (aref subj_V8 count)))
;; Vision Cone
      (def posVC (matrix-trans (aref subj_VC count)))
;; Left Palm array
      (def posLP (matrix-trans (aref subj_LP count)))
;; Left foreArm array
      (def posLA (matrix-trans (aref subj_LA count)))
;; HeaD array
      (def posHD (matrix-trans (aref subj_HD count)))
;; Lower Body array
      (def posLB (matrix-trans (aref subj_LB count)))
;; Left Leg array
      (def posLL (matrix-trans (aref subj_LL count)))
;; Right Leg array
      (def posRL (matrix-trans (aref subj_RL count)))

;;0  SIMULATION TIME
   (def countn (+ (/ (* count 2) 60) .03))
     (format cf "~5,2F "countn)
;;1 CALCULATE DISTANCE BETWEEN LOWER BODY AND SITE 5
(format cf "~4,0F\ " (sqrt (+ (* (- (aref posLB 0)  (aref posS5 0))
(- (aref posLB 0)  (aref posS5 0))) (* (- (aref posLB 1)  (aref posS5 1))
(- (aref posLB 1)  (aref posS5 1))) (* (- (aref posLB 2)  (aref posS5 2))
(- (aref posLB 2)  (aref posS5 2))))))
;; collision detection with viewcone and boom
;;2 CALCULATE DISTANCE BETWEEN V1 AND  SITE 6
(format cf "~5,0F\ " (- 72.88 (sqrt (+ (* (- (aref posV1 0)  (aref posS6 0))
(- (aref posV1 0)  (aref posS6 0))) (* (- (aref posV1 1)  (aref posS6 1))
(- (aref posV1 1)  (aref posS6 1))) (* (- (aref posV1 2)  (aref posS6 2))
(- (aref posV1 2)  (aref posS6 2)))))))
;;3 CALCULATE DISTANCE BETWEEN V2 AND  SITE 6
(format cf "~5,0F\ " (- 72.88 (sqrt (+ (* (- (aref posV2 0)  (aref posS6 0))
(- (aref posV2 0)  (aref posS6 0))) (* (- (aref posV2 1)  (aref posS6 1))
(- (aref posV2 1)  (aref posS6 1))) (* (- (aref posV2 2)  (aref posS6 2))
(- (aref posV2 2)  (aref posS6 2)))))))
;;4 CALCULATE DISTANCE BETWEEN V3 AND  SITE 6
(format cf "~5,0F\ " (- 103.66 (sqrt (+ (* (- (aref posV3 0)  (aref posS6 0))
(- (aref posV3 0)  (aref posS6 0))) (* (- (aref posV3 1)  (aref posS6 1))
(- (aref posV3 1)  (aref posS6 1))) (* (- (aref posV3 2)  (aref posS6 2))
(- (aref posV3 2)  (aref posS6 2)))))))
;;5 CALCULATE DISTANCE BETWEEN V4 AND  SITE 6
(format cf "~5,0F\ " (- 103.66 (sqrt (+ (* (- (aref posV4 0)  (aref posS6 0))
(- (aref posV4 0)  (aref posS6 0))) (* (- (aref posV4 1)  (aref posS6 1))
(- (aref posV4 1)  (aref posS6 1))) (* (- (aref posV4 2)  (aref posS6 2))
(- (aref posV4 2)  (aref posS6 2)))))))
;;6 CALCULATE DISTANCE BETWEEN V5 AND  SITE 6
(format cf "~5,0F\ " (- 131.97 (sqrt (+ (* (- (aref posV5 0)  (aref posS6 0))
(- (aref posV5 0)  (aref posS6 0))) (* (- (aref posV5 1)  (aref posS6 1))
(- (aref posV5 1)  (aref posS6 1))) (* (- (aref posV5 2)  (aref posS6 2))
(- (aref posV5 2)  (aref posS6 2)))))))
;;7 CALCULATE DISTANCE BETWEEN V6 AND  SITE 6
```

```
(format cf "~5,0F\ " (- 131.97 (sqrt (+ (* (- (aref posV6 0)  (aref posS6 0))
 (- (aref posV6 0)  (aref posS6 0))) (* (- (aref posV6 1)  (aref posS6 1))
 (- (aref posV6 1)  (aref posS6 1))) (* (- (aref posV6 2)  (aref posS6 2))
 (- (aref posV6 2)  (aref posS6 2)))))))
;;8 CALCULATE DISTANCE BETWEEN V7 AND  SITE 6
(format cf "~5,0F\ " (- 102.0 (sqrt (+ (* (- (aref posV7 0)  (aref posS6 0))
 (- (aref posV7 0)  (aref posS6 0))) (* (- (aref posV7 1)  (aref posS6 1))
 (- (aref posV7 1)  (aref posS6 1))) (* (- (aref posV7 2)  (aref posS6 2))
 (- (aref posV7 2)  (aref posS6 2)))))))
;;9 CALCULATE DISTANCE BETWEEN V8 AND  SITE 6
(format cf "~5,0F\ " (- 102.0 (sqrt (+ (* (- (aref posV8 0)  (aref posS6 0))
 (- (aref posV8 0)  (aref posS6 0))) (* (- (aref posV8 1)  (aref posS6 1))
 (- (aref posV8 1)  (aref posS6 1))) (* (- (aref posV8 2)  (aref posS6 2))
 (- (aref posV8 2)  (aref posS6 2)))))))
;;10  CALCULATE DISTANCE BETWEEN BOOM AND REF PT SITE 5...BOOM MOTION INDICATOR
(format cf "~5,0F\ " (sqrt (+ (* (- (aref posS1 0)  (aref posS5 0))
 (- (aref posS1 0)  (aref posS5 0))) (* (- (aref posS1 1)  (aref posS5 1))
 (- (aref posS1 1)  (aref posS5 1))) (* (- (aref posS1 2)  (aref posS5 2))
 (- (aref posS1 2)  (aref posS5 2))))))

;;;; COLLISION DETECTIONS WITH BODY PARTS
;;11     LOWER PALM WITH BOOM ARM
   (format cf "~4,0F\ " (aref subj_collLPB count))
;;12     LOWER PALM WITH DRILL HEAD
   (format cf "~4,0F\ " (aref subj_collLPD count))
;;13     LOWER ARM WITH BOOM ARM
   (format cf "~4,0F\ " (aref subj_collLAB count))
;;14     LOWER ARM WITH DRILL HEAD
   (format cf "~4,0F\ " (aref subj_collLAD count))
;;15     LEFT LEG WITH BOOM ARM
   (format cf "~4,0F\ " (aref subj_collLLB count))
;;16     LEFT LEG WITH DRILL HEAD
   (format cf "~4,0F\ " (aref subj_collLLD count))
;;17     RIGHT LEG WITH BOOM ARM
   (format cf "~4,0F\ " (aref subj_collRLB count))
;;18     RIGHT LEG WITH DRILL HEAD
   (format cf "~4,0F\ " (aref subj_collRLD count))
;;19     HEAD WITH BOOM ARM
   (format cf "~4,0F\n" (aref subj_collHDB count))
 )
 (format cf "  \n")    ;end loop to print data
 (close cf)
)
```

www.ingramcontent.com/pod-product-compliance
Lightning Source LLC
Chambersburg PA
CBHW081814170526
45167CB00008B/3436